SpringerBriefs in Molecular Science

History of Chemistry

Series editor

Seth C. Rasmussen, Department of Chemistry and Biochemistry, North Dakota State University, Fargo, ND, USA

More information about this series at http://www.springer.com/series/10127

Seth C. Rasmussen

Acetylene and Its Polymers

150+ Years of History

 Springer

Seth C. Rasmussen
Department of Chemistry and Biochemistry
North Dakota State University
Fargo, ND, USA

ISSN 2191-5407   ISSN 2191-5415   (electronic)
SpringerBriefs in Molecular Science
ISSN 2212-991X
SpringerBriefs in History of Chemistry
ISBN 978-3-319-95488-2   ISBN 978-3-319-95489-9   (eBook)
https://doi.org/10.1007/978-3-319-95489-9

Library of Congress Control Number: 2018947476

This Springer imprint is published by the registered company Springer International Publishing AG part of Springer Nature
The registered company address is: Gewerbestrasse 11, 6330 Cham, Switzerland

Acknowledgements

I would first and foremost like to thank the Department of Chemistry and Biochemistry at North Dakota State University (NDSU) for supporting my continuing efforts in the history of chemistry. In addition, I need to acknowledge the American Chemical Society's Division of the History of Chemistry (HIST) in providing the environment and encouragement which allowed the development of my initial historical interests into my current active research and contributions in the history of science. I also must acknowledge the Australian-American Fulbright Commission who supported my stay at the University of Newcastle as a Fulbright senior scholar for the spring of 2018, during which large sections of the current volume were written.

The strength of this particular volume was due to the contributions of a number of friends and colleagues. As such, I would like to thank Prof. Alan Rocke (Case Western Reserve University) for helpful discussions on the issues of correctly determining molecular formulas of organic species during the nineteenth century, Dr. Vera V. Mainz (University of Illinois) and Elsa B. Atson (Science History Institute) for assistance with tracking down sources on Paul and Arnould Thenard, Dr. Reggie Hudson (NASA Astrochemistry Laboratory) for helpful discussions and feedback on cuprene, as well as for providing the color image of a cuprene film, Prof. Yasu Furukawa (Nihon University) for assistance in finding biographical information on Hatano and Ikeda, and Prof. Dennis Cooley (NDSU) for help with philosophy sources relating to the nature of discovery. Most critically, I would like to thank Prof. John Reynolds (Georgia Tech) for initially bringing the work of Smith and Berets to my attention, which initiated much of my detailed re-evaluation of the history of polyacetylene and Dr. Choon Do for his help in locating Dr. Hyung Chick Pyun and his assistance in collecting much of his biographical material.

In addition, I need to acknowledge the Interlibrary Loan Department of NDSU, who went out of their way to track down many elusive and somewhat obscure sources, as well as my brother, Dr. Kent A. Rasmussen, for continued assistance with translations. I could not imagine doing historical work without both of these critical resources. I also need to thank the following current and former members of my materials research group for reading various drafts of this manuscript and providing critical feedback: Prof. Christopher L. Heth (Minot State University), Kristine L. Konkol, Even W. Culver, Dr. Eric J. Uzelac (Marvin Windows), and Trent E. Anderson.

Finally, and perhaps most importantly, I must express my continued thanks to Elizabeth Hawkins at Springer, without whom this growing series of historical volumes would not have become a reality, as well as Sofia Costa who now oversees the series as its managing editor at Springer.

Contents

About the Author

Seth C. Rasmussen is a professor of Chemistry at North Dakota State University (NDSU), Fargo (seth.rasmussen@ndsu.edu). A native of the Seattle area, he received his B.S. in Chemistry from Washington State University in 1990, before continuing his graduate studies at Clemson University under the guidance of Prof. John D. Petersen. After completing his Ph.D. in Inorganic Chemistry in 1994, he moved to the University of Oregon to study conjugated organic polymers as a postdoctoral associate under Prof. James E. Hutchison. He then accepted a teaching position at the University of Oregon in 1997, before moving to join the faculty at NDSU in 1999. Attaining the rank of full professor in 2012, he also spent the spring of 2018 as a Fulbright senior scholar and a visiting professor at the Centre for Organic Electronics of the University of Newcastle, Australia.

Active in the fields of both materials chemistry and the history of chemistry, his research interests include the design and synthesis of conjugated materials, photovoltaics (solar cells), NIR photodetectors, organic light-emitting diodes, the history of materials, chemical technology in antiquity, and the application of history to chemical education. As both author and editor, he has contributed to books in both materials and history and has published more than 95 research papers and chapters. He is a member of various international professional societies including the American Chemical Society, Materials Research Society, Alpha Chi Sigma, American Nano Society, Society for the History of Alchemy and Chemistry, History of Science Society, and the International History, Philosophy & Science Teaching Group.

He served as the program chair for the History of Chemistry (HIST) Division of the American Chemical Society from 2008 to 2017 and continues to serve the division as a member-at-large of its executive committee. In addition, he currently serves as the series editor for the *SpringerBriefs in Molecular Science: History of Chemistry* book series, as an editor of the journal *Cogent Chemistry*, and on the advisory board for the journal *Substantia: An International Journal of the History of Chemistry*.

Abstract

The polymerization of acetylene is most commonly associated with the production of polyacetylene, which was found to be conductive when treated with oxidizing agents such as Br_2 or I_2 in the mid-to-late 1970s. In fact, under the right conditions, oxidized polyacetylenes can exhibit conductivities into the metallic regime, thus providing the first example of an organic polymer exhibiting metallic conductivity. As a consequence, the 2000 Nobel Prize in Chemistry was awarded to Hideki Shirakawa, Alan MacDiarmid, and Alan Heeger for this pioneering research, the award citation reading "for the discovery and development of electrically conductive polymers." Because of this, most view polyacetylene, as well as conducting polymers in general, to originate in the 1970s, neither of which is historically correct. Although true polyacetylene was not successfully produced until the 1950s by Giulio Natta, the polymerization of acetylene dates back to 1866 with the work of Marcelin Berthelot. These initial efforts were continued by a range of scientists to produce a polymeric material collectively given the name cuprene in 1900 by Paul Sabatier. Between the initial cuprene studies and the production of true polyacetylene, two related materials were also studied, usually referred to as polyenes and polyvinylenes. Although both of these materials could be thought of as forms of polyacetylenes, neither was actually generated from the direct polymerization of acetylene. For the first time, this volume aims to present a historical overview of the development of these materials, beginning with the initial discovery of acetylene in 1836 and continuing up through the 2000 Nobel Prize in Chemistry. In the process, the reader will hopefully gain insight into the fact that polyacetylene and conducting organic polymers have a much longer history than commonly believed and involved the work of a significant number of Nobel laureates.

Keywords Acetylene · Polyacetylene · Cuprene · Polyvinylene
Polyene · Conducting polymers

Chapter 1
Introduction

Gaseous molecules were central to the development of modern chemistry, with most of the fundamental work of luminaries such as Joseph Black (1728–1799), Antoine Lavoisier (1743–1794), and John Dalton (1766–1844) all involving gases [1]. This was partly because gases could be easily prepared in a pure condition and partly because the mathematical relationships between the volumes of gaseous reactants and products are simple. The discovery of gases as a physical state is typically credited to Johannes Baptista Van Helmont (1579–1644)[1] (Fig. 1.1) who was the first to distinguish gaseous products from common air in the early 17th century [2, 4, 5].

Van Helmont could distinguish various gases on the basis of such characteristics as color, odor, or taste and thus concluded that they were not simple air. All attempts to isolate these gases failed, however, as his receivers always burst, thus prohibiting him

---

[1] Johannes Baptista Van Helmont was born in Brussels in 1579 [1–3] and is considered by some as the most prominent chemist of the first half of the 17th century [3, 4], although his contributions to medicine are more significant than those of chemistry. Even though his name is given as Johannes in the frontispiece shown in Fig. 1.1, it has also been reported by various authors as Johann [3, 4], Joan [5, 6], Jean, and Jan. To avoid this confusion, some authors only give his initial. Coming from a noble family, Van Helmont studied at the University of Louvain until 1594 [2–4]. Although educated in the conventional classical courses, he took no degree as he considered academic honors a mere vanity [2, 3]. His interests then led him to the medical profession, and he took a M.D. at Louvain in either 1599 [2, 4] or 1609 [3]. He died on December 30, 1644, either in Brussels or Vilvorde [3].

© The Author(s) 2018
S. C. Rasmussen, *Acetylene and Its Polymers*, SpringerBriefs in Molecular Science,
https://doi.org/10.1007/978-3-319-95489-9_1

Fig. 1.1 Johannes Baptista
Van Helmont (1579–1644)
(Adapted from the
frontispiece to Van
Helmont's *Oriatrike* (1662))

from obtaining pure gases [2, 3, 5]. The collection and isolation of gases would have
to wait until the invention of the pneumatic trough by Stephen Hales (1677–1761)[2]
(Fig. 1.2). In its original form, this was simply a tube leading from the area of heating
to an inverted bottle full of water [9, 10]. When the reaction mixture was heated,
any gas generated passed through the tube and bubbled up through the water into
the inverted bottle. As the gas accumulated at the top, it pushed the water out the
bottom. When all the water was evacuated, the full bottle of gas was stoppered, thus
allowing the quantitative collection of gases.

In general, gases are typically rather small molecules with very weak intermolecular forces, such that the sum of the intermolecular forces involved is less than the
available kinetic energy at room temperature. It should be pointed out that in modern
usage, gases are differentiated from vapors, with the later representing the gaseous
form of species normally a liquid or solid at room temperature.

---

[2] Stephen Hales was born September 7, 1677 [7, 8], at Bekesbourne in Kent, the sixth son of Thomas
and Mary Hales [7]. He entered Bennet College, Cambridge, in 1696, at the age of nineteen [7, 8].
In ca. 1702, he took a Bachelor of Arts degree [7]. He was then ordained a deacon, elected a fellow,
and granted a Master of Arts degree in 1703 [7, 8]. He continued studies until 1708–1709 [7]. In
ca. 1710, he was made perpetual curate of Teddington in Middlesex [7, 9] and became a Bachelor
of Divinity in 1711 [7]. He later took the Doctor of Divinity at the University of Oxford [8] and
refused a canonry of Winsor so that he could devote himself to his experimental pursuits [9]. He
was elected a Fellow of the Royal Society in 1718 [8] and was awarded the Copley Medal in 1739
[7, 8]. He died January 4, 1761, after a short illness [7].

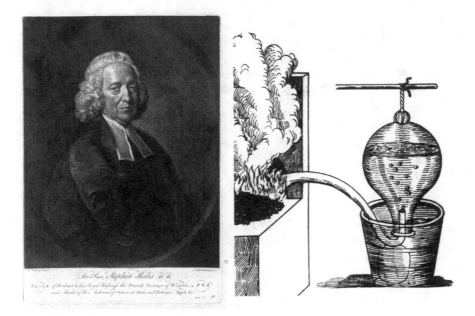

Fig. 1.2 Stephen Hales (1677–1761) at age 82 (Mezzotint by J. McArdell after T. Hudson, Courtesy of Wellcome Library, London, under Creative Commons Attribution only licence CC BY 4.0) and his pneumatic trough

A subclass of such gaseous molecules are gaseous hydrocarbons, that is, organic compounds consisting entirely of hydrogen and carbon which are gases at room temperature and pressure. Common examples of gaseous hydrocarbons are given in Fig. 1.3, the simplest of which is methane. Such hydrocarbons can be further subdivided into saturated hydrocarbons (e.g. alkanes) and unsaturated hydrocarbons (e.g. alkenes and alkynes). The two simplest examples of this latter class are ethylene and acetylene, the second of which is the focus of the current volume.

Of course, the systematic names given in Fig. 1.3 were not in use until their introduction by August Wilhelm von Hofmann (1818–1892)[3] (Fig. 1.4) in 1866, when he proposed a new system of nomenclature for hydrocarbons in order to reduce confusion in their discussion [13]. In his proposed system, saturated hydrocarbons were

---

[3] August Wilhelm Hofmann was born on April 8, 1818 in Giessen. He began studying philosophy and law in University of Giessen in 1836 [11, 12]. His course of study also included the fundamentals of chemistry, which exposed him to the laboratory of Justus von Liebig (1803–1873) [12]. Hofmann eventually turned to chemistry and was awarded his Ph.D. in 1841. By 1843, Hofmann had become Liebig's assistant [11, 12]. He then became a Privatdozent in Bonn in the spring of 1845 [12]. Shortly thereafter, though, Hofmann was offered the directorship of the Royal College of Chemistry in London. To facilitate the offer, he was appointed assistant professor at Bonn and granted a leave of absence [12]. Thus, he moved to England in 1845 with the plan to return to Bonn after two years. He became a Fellow of the Royal Society in 1851 [11]. He ultimately returned to Germany in 1865 to succeed Eilhard Mitscherlich (1794–1863) at Berlin. It was after this return that he also become von Hofmann [11].

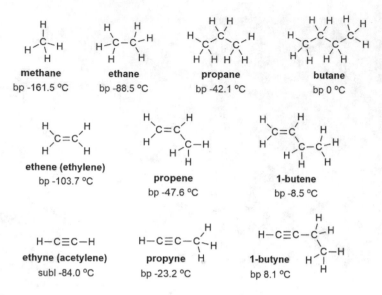

Fig. 1.3 Common gaseous hydrocarbons

Fig. 1.4 August Wilhelm
von Hofmann (1818–1892)
(Edgar Fahs Smith Memorial
Collection. Kislak Center for
Special Collections, Rare
Books and Manuscripts.
University of Pennsylvania)

given the suffix -*ane*, along with a prefix consisting of "the first syllable of the Latin numeral corresponding to the number of carbon atoms in the molecule." Unsaturated hydrocarbons where named in the same fashion, but using the suffixes -*ene* or -*ine* to indicate the removal of two or four hydrogens respectively. Hofmann's system eventually became a starting point for the International Commission of Chemical Nomenclature during the development of the Geneva Nomenclature of 1892 [14]. This system still remains at the core of modern IUPAC nomenclature, although with some modifications over time.[4] Most critical to the topic of the current volume was the change of the suffix -*ine* to -*yne* to indicate hydrocarbons containing triple bonds, which occurred as part of the modifications adopted during the Liege Nomenclature of 1930 [15]. Prior to the development of this formal nomenclature, however, the species in Fig. 1.3 were known by a variety of names. By 1892, the early names ethylene and acetylene were so deeply entrenched in the literature that these terms continued to be used and are still more commonly used than their formal IUPAC names.

1.1 Saturated Gaseous Hydrocarbons

The simplest saturated hydrocarbon gas is methane (CH_4), which is the main constituent of natural gas. As such, it was also the first of these gaseous hydrocarbons to be studied. The gas was known by a number of different terms, one of which was *firedamp*, that described a number of flammable gases found in coal mines, particularly methane. Explosions in mines have been recorded back to at least the 17th century [16]. However, the most common early term for methane was *marsh gas*, a flammable gas produced naturally within some geographical marshes, swamps, and bogs. In November of 1776, Alessandro Volta (1745–1827)[5] (Fig. 1.5) studied the combustion properties of this "flammable air" from samples he had collected

---

[4]In Hofmann's original nomenclature, the modern *butane* was given the name *quartane* [13].

[5]Alessandro Volta was born in Camnago on February 18, 1745 [17]. He began his career as a teacher of physics in the gymnasium at Como in 1774 [18]. He then became professor of physics at the University of Pavia in 1779 [17, 18], where he began his electrical research for which he is so well known. He was then elected Rector at Pavia in 1785 [17]. From 1791, he worked entirely on galvanism, with his voltaic pile (the modern battery) reported in 1800. In 1801, he was summoned to Paris by Napoleon and was presented with a gold medal for his work. He became a Fellow of the Royal Society in 1791 and an Officer of the Legion of Honour in 1802. He retired from his professorship at Pavia in 1803, but became Director of the Philosophical Faculty at Padua in 1815. In 1819, he retired from all offices [18] and returned to Como, where he died on March 5, 1827 [17].

Fig. 1.5 Alessandro Volta (1745–1827) (left) and John Dalton (1766–1844) (right) (Edgar Fahs Smith Memorial Collection. Kislak Center for Special Collections, Rare Books and Manuscripts. University of Pennsylvania)

in the marshes of Lake Maggiore [19]. Later, in 1808, John Dalton[6] (Fig. 1.5) also studied marsh gas, which he called *carburetted hydrogen* [22]. He analyzed the gas via sparking and combustion, concluding the gas was comprised of only carbon and hydrogen, giving the formula CH_2.[7] He also determined that this gas was the primary component of *coal gas*, which could be obtained from the distillation of pitcoal [22]. Marcelin Berthelot (1827–1907), who will be discussed in much more detail in the following chapters, synthesized methane in 1856 by passing a stream of gaseous carbon disulphide and hydrogen sulphide over red-hot copper [23, 24]. This resulted in a mixture of hydrogen, marsh gas (methane), olefiant gas (ethylene), and a trace of naphthalene, from which the methane could be isolated via solvent extraction [23].

---

[6]John Dalton was born September 5, 1766 in a tiny hamlet of Eaglesfield [20], the son of a poor weaver and farmer [21]. He was informally educated until age 12 [20, 21], after which he opened a village school and began teaching himself [20]. In 1781, Dalton and his brother moved to Kendal to teach at their cousin's Quaker school. The brothers took over the school in 1785 [20]. In 1793, he become a professor of natural philosophy at New College in Manchester [20, 21]. Dalton resigned from New College in 1800, after which he supported himself by private teaching and occasional lecturing [20]. He was elected a corresponding member of the Academie des Sciences in 1816 and raised to a Foreign Associate in 1830. He became a Fellow of the Royal Society in 1822 and he became the first recipient of the Royal Medal in 1826. He died on July 27, 1844 [20].

[7]Many atomic weights were not yet formalized in 1810, including carbon. Dalton came to the formula CH_2, rather than CH_4, because he used the value of 5.4 for the atomic weight of carbon [22].

Fig. 1.6 Michael Faraday (1791–1867) (Edgar Fahs Smith Memorial Collection. Kislak Center for Special Collections, Rare Books and Manuscripts. University of Pennsylvania)

Ethane may have been prepared as early as 1833, during the electrolysis of potassium acetate solutions by Michael Faraday (1791–1867)[8] (Fig. 1.6) [27]. When concentrated solutions were utilized, the products consisted of carbonic oxide and carbonic acid at the anode and hydrogen at the cathode. In contrast, when dilute solutions were used, he observed less production of carbonic oxide at the anode and the production of a hydrocarbon gas at the cathode, rather than hydrogen. Believing the hydrocarbon to be methane, however, he did not investigate it further [27].

---

[8]Michael Faraday was born September 22, 1791 in Newington in Surrey [25]. His father suffered from ill-health and the family was thus fairly poor. Faraday was apprenticed to a local bookseller and bookbinder at age 14 [25]. After attending lectures of Humphry Davy in 1812, he persuaded Davy to take him on as an assistant [25] and entered the Royal Institution in 1813, at age 21, as Humphry Davy's laboratory assistant [26]. After a leave of absence to accompany Davy and his wife on a continental tour, he returned as assistant to William Brande, Davy's successor as professor of chemistry. In 1821 Faraday was appointed Superintendent of the House and Laboratory in 1821 and he became Director of the Laboratory in 1825 [25, 26]. He was finally appointed Fullerian professor of chemistry in 1834, at age 42 [26]. He died August 25, 1867 [25].

Fig. 1.7 Edward Frankland (1825–1899) (left) and Hermann Kolbe (1818–1884) (right) (Edgar Fahs Smith Memorial Collection. Kislak Center for Special Collections, Rare Books and Manuscripts. University of Pennsylvania)

In an effort to vindicate the radical theory[9] of the time, Edward Frankland (1825–1899)[10] and Hermann Kolbe (1818–1884)[11] (Fig. 1.7) reacted *cyanide of ethyl* (propionitrile) with potassium in an attempt to isolate the ethyl radical beginning in 1848 [31, 32]. This reaction generated a colorless gas that could not be condensed at a temperature of 18 °C. Analysis of the gas, however, led to the conclusion that what they had isolated was the methyl radical [32]. Believing that the methyl radical was produced by decomposition of the ethyl radical, Franklin then attempted to use more

---

[9]The original meaning of the term *radical* was very different than the modern definition and was not defined by an unpaired electron. In its use in the 19th century, a radical referred to the root base of a series of compounds (such as a methyl or ethyl radical) that could be isolated as a discrete compound.

[10]Edward Frankland was initially apprenticed to a pharmacist in Lancaster, but it was a doctor that befriended him who taught him Dalton's atomic theory, gave him facilities for laboratory work, and advised him to study with Lyon Playfar (1818–1898) at the College of Engineers in Putney [28]. He became Playfar's chief assistant in 1847 and it was there that he met Kolbe. That same year the two worked for some time in Robert Bunsen's laboratory in Marburg, after which Frankland returned to Marburg the next year and spent time with Justus von Liebig in Giessen in 1849. Frankland succeeded Playfar at Putney and then became professor of chemistry at Owens College, Manchester in 1851. He then became professor at St. Bartholomew's Hospital in 1857, the Royal Institution in 1863, and finally Hofmann's successor at the Royal School of Mines in 1865 [28, 29]. He resigned his professorship in 1885 [28].

[11]Hermann Kolbe was the oldest of 15 children, the son of a Lutheran pastor [30]. He studied under Friedrich Wöhler (1800–1882) in 1838, was assistant to Bunsen in 1842, and Playfair in 1845. In 1847, he returned to Marburg and succeeded Bunsen as professor there in 1851. He moved to Leipzig in 1865 where he had a lab built for 132 students. Besides being a talented experimenter, Kolbe had a reputation as a very successful teacher [30].

gentle conditions to generate the ethyl radical. Thus, in 1849, he studied the reaction of ethyl iodide with zinc and concluded that the gas isolated from this reaction was a mixture of the methyl and ethyl radicals [33, 34]. By 1851, however, Frankland realized that at least a portion of the gas generated by all of these methods was in fact *hydride of ethyl* (ethane) [35]. It is now known that what he believed to be the ethyl radical was in fact butane [36].

Propane was first prepared by Berthelot in 1857. By heating propylene dibromide (1,2-dibromopropane) with copper, potassium iodide, and water at 275 °C, he produced a mixture of propylene and *hydrure de propyle* (hydride of propyl, i.e. propane) [37]. He was able to produce similar results from dibromopropene and tribromopropane as well. Lastly, he was able to treat glycerine with hydrochloric acid to give trichloropropane, which when treated as above, also produced propane. Thus, in this way, he was able to generate propane from glycerine [37]. Berthelot then returned to this chemistry in 1867 to show that propane could be produced via the heating of allyl iodide, acetone, or glycerine with hydroiodic acid at 275 °C [38].

These saturated gaseous hydrocarbons are primarily utilized as fuels, one of the earliest examples of which was gas lighting (Fig. 1.8). Such lighting consisted of the production of artificial light from the combustion of gaseous fuels such as methane, propane, butane, ethylene, or acetylene and dates back to the late 18th century. Early fuel and lighting gas did not consist of a single pure gaseous species, but a mixture of gases obtained from the pyrolysis of carbon species such as coal or wood. Methane and ethane are gaseous at ambient temperatures and cannot be readily liquefied by pressure alone. Propane, however, is easily liquefied, and exists in modern commercial propane bottles mostly as a liquid. In comparison, butane is so easily liquefied that it provides a safe, volatile fuel for small pocket lighters.

1.2 Unsaturated Gaseous Hydrocarbons

The first of the gaseous hydrocarbons to be synthesized was ethylene and, for a period of time, *olefiant gas* (ethylene or ethene) and marsh gas (methane) were the only two gaseous hydrocarbons known [39]. Ethylene may have been obtained by Johann Becher (1635–1682) as early as 1669 by heating alcohol with sulfuric acid. In a similar manner, Jan Ingen-Housz (1730–1799)[12] (Fig. 1.9) reported in 1779 that during a visit to Amsterdam in 1777, he had observed two gentlemen (Aeneae

---

[12] Jan Ingen-Housz was born in Breda, Netherlands, on December 8, 1730 [40, 41]. He was educated at the Breda Latin School until age 16. He then began studying at the University of Louvain, where he received the MD degree in 1753 at the age of 22 [40, 41]. Afterwards, he continued studies for several years at the universities of Leiden, Paris and Edinburgh [40, 41], before returning to Breda in 1756 to begin practicing medicine [41]. In 1765, he moved to London where he learned the technique of inoculation against smallpox, using the live virus. He traveled to Vienna in 1768 to inoculate the Royal Family [40, 41]. His success led to an appointment as the Court Counsellor and Personal Physician to the Imperial Family [41]. Ingen-Housz is also credited with the discovery of photosynthesis [40]. Ingen-Housz died on September 7, 1799 [40, 41].

Fig. 1.8 Gas Light on Charles Street in Boston (Photo by Ben Wildeboer, courtesy Creative Commons Attribution-ShareAlike 2.0 Generic license CC BY-SA 2.0)

and Cuthbertson) prepare an inflammable air from the heating of equal quantities of sulfuric acid and alcohol [42]. The reaction produced large amounts of white vapor that became clear when collected over water. The gas had an ether smell, was reportedly heavier than common air, and highly flammable when mixed with even a tenth part common air. Others too, had also observed the production of a flammable gas by passing alcohol vapor over various hot metals [43].

The first detailed study of the gaseous products obtained from these various reactions of alcohol, however, was reported in 1795 [44] by the group known collectively as "the Dutch chemists" [43]. This group was made up of Johann Rudolph Deimann (1743–1808), Adrien Paets van Troostwyk (1752–1837), Nicolas Bondt (1765–1796), and Anthoni Lauwerenburgh (1758–1820) [43]. They began their study by optimizing the production of the gas, finding that the yield of product could be maximized by using a ratio of four parts of concentrated sulfuric acid to one part of alcohol by weight [44]. The gas was then easily purified by washing with water. The gas was described as having a very unpleasant smell, providing it was washed well with water to remove small quantities of ether byproduct, and was determined to have a specific gravity of 0.905 (relative to a value of 1.000 for common air).[13] Analysis of

---

[13]The modern density value for ethylene is 1.178×10^{-3} g/mL, while the density of air is 1.2041×10^{-3} g/mL at 20 °C. Thus, using air as a standard, this would give a specific gravity of 0.978 for ethylene, in reasonable agreement with the value of the Dutch chemists. However, this is not in agreement with the observation by Ingen-Housz that the gas he observed was heavier than air.

Fig. 1.9 Jan Ingen-Housz
(1730–1799) (Edgar Fahs
Smith Memorial Collection.
Kislak Center for Special
Collections, Rare Books and
Manuscripts. University of
Pennsylvania)

the gas led to the conclusion that it was comprised of only carbon and hydrogen and
they gave it the name *gaz hydrogène carboné huileux* [40]. Their discovery of this
gas was communicated to the Paris Institut in March of 1796, which was then communicated by Fourcroy in August 1796, who referred to the gas as *gaz olefiant* [43].

Acetylene, the topic of focus for the current volume, was first produced in 1836
by Edmund Davy (1785–1857), who called it *bicarburet of hydrogen* [45, 46]. At
this point, only marsh gas and olefiant gas were known and thus it was only the
third known hydrocarbon gas. This discovery did not have the initial impact that
ethylene did, however, and thus Davy's work was forgotten for a while until acetylene
was rediscovered by Berthelot in 1860 [47]. It was Berthelot that gave the gas the
name *acétylène*. This history will be revisited in much greater detail in the following
chapter.

Propylene was then discovered by Captain John W. Reynolds[14] in 1850 [49], who
decomposed amyl alcohol (pentanol, $C_5H_{12}O$) by passing its vapor through a red-
hot, potash-glass tube. Based upon the dehydration of ethanol to yield ethylene, this
would be expected to yield pentene, yet the high temperature causes decomposition

---

[14]Little is known about Reynolds other than the fact that he was a student of Hofmann [48].

of the higher hydrocarbon to give propylene along with other byproducts. The gas produced was highly dependent on the temperature, with the highest temperatures only yielding methane. In contrast, if the temperature was too low, the amyl alcohol just distilled and was recovered. As a result of this strong temperature dependence, the results varied considerably with each attempt and always resulted in a mixture of various products [49]. Because of this, the gaseous product mixture was collected and then the propylene fraction converted to either its dibromide or dichloride by reaction with the corresponding dihalide (liquid Br_2 or gaseous Cl_2). As propylene was the only gaseous product to react with the dihalides and resulted in a liquid product, the corresponding halide derivatives could thus be isolated as pure materials. Berthelot was then able to convert the dihalides back to a mixture of propylene and propane in 1857 as described above [37].

As with the previously discussed saturated hydrocarbon gases, both ethylene and acetylene were components of gases used as fuels and in gas lighting. Ethylene was also used as an anesthetic up through the 1940s [50]. However, for the focus of the current volume, the most important application of such unsaturated hydrocarbons was as monomeric precursors for polymeric materials.

1.3 Polymerization of Unsaturated Gaseous Hydrocarbons

Small molecules with double or triple bonds, such as unsaturated gaseous hydro-carbons, can be used as precursors for the generation of macromolecular species, commonly referred to as polymers. It is important to remember, however, that the meaning of the word polymer has changed over time and that it predates the concept of the macromolecule introduced by Hermann Staudinger (1881–1965)[15] in the 1920s [53–55]. The term *polymer* originates with Jöns Jacob Berzelius (1779–1848)[16]

---

[15]Hermann Staudinger was born in Worms, Germany on March 23, 1881 [51]. He studied at Darmstadt, Munich, and Halle. After taking his doctorate in 1903, he worked under Johannes Thiele (1865–1918) at Strasbourg until 1907 [51]. He was then made an associate professor at the Karlsruhe Technische Hochschule (Karlsruhe Institute of Technology) before succeeding Richard Willstatter (1872–1942) as professor of organic chemistry at the Federal Institute of Technology in Zurich (ETH Zurich) in 1912 [51, 52]. He then moved to the University of Freiburg in 1925 [51, 52]. He was awarded the Nobel Prize in 1953 "for his discoveries in the field of macromolecular chemistry." He retired in 1951 and died September 8, 1965 [51].

[16]Jöns Jacob Berzelius (1779–1848) was born in a small Swedish town in East Gothland. As both of his parents died when he was young, he was raised by his stepfather Anders Ekmarck. He finished school in 1796, after which he entered the University of Uppsala as a medical student. Due to lack of means, however, he was forced to leave and became a private tutor until he won a small scholarship in 1798. He then reentered the University and graduated with a dissertation on mineral water [56]. He completed his M.D. in 1802 with a thesis on the medical applications of galvanism and was appointed reader in chemistry at the Carlberg Military Academy in 1806. He was appointed professor of medicine and pharmacy the following year at the School of Surgery in Stockholm, where he had a modest laboratory [56]. He was elected a member of the Swedish Academy of Sciences in 1808 and became a joint secretary in 1818. He resigned his professorship in 1832, but continued to be active in chemical discussions until his death in 1848 [56].

Fig. 1.10 Jöns Jacob
Berzelius (1779–1848)
(Edgar Fahs Smith Memorial
Collection. Kislak Center for
Special Collections, Rare
Books and Manuscripts.
University of Pennsylvania)

(Fig. 1.10), who introduced the term *polymeric* in 1832 [54, 55, 57]. In its introduction by Berzelius, a polymer was a type of isomeric form in which "the relative number of atoms is equal, but the absolute number is unequal" [57]. Thus, by this definition, benzene (C_6H_6) would be a polymer of acetylene (C_2H_2). Of course, most homopolymeric macromolecules would also fit this definition, but neither macromolecular nature nor high molecular weight was a factor in the original use of the term polymer. Over time, the term evolved into the modern meaning, but when discussing older literature, particularly that of the 19th century, the use of the term polymer alone does not ensure that the product described is a long-chained macromolecular species and other indicators must be utilized to determine the true nature of the species of interest.

The conversion of unsaturated hydrocarbons to macromolecular species occurs via addition polymerization [58, 59]. As shown in Fig. 1.11, addition polymerizataion requires an initiation step, which forms a reactive species from which the polymer chain grows via propagation. This initiation step can involve a number of different types of initiators, including free radical, anionic, or cationic species, as well as coordination to an active metal compound (i.e. Ziegler-Natta catalysts). In each of these cases, the propagating species would consist of a reactive center of the same type as the initiator (i.e. a free radical initiator generates a free radical propagating species).

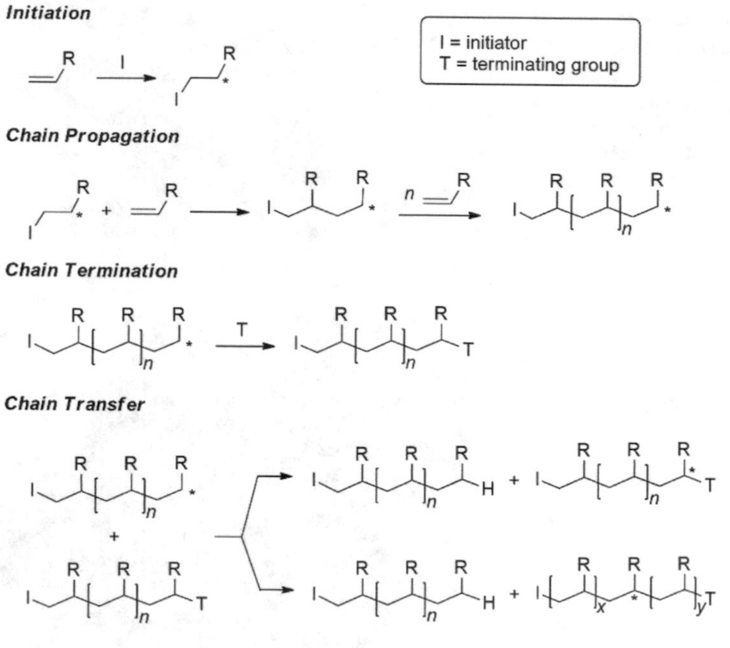

Initiation

Chain Propagation

Chain Termination

Chain Transfer

Fig. 1.11 General mechanism for addition polymerization

In some cases, initiation of addition polymerization can also be accomplished via heat or light. For photoinitiated polymerization, the unsaturated unit to undergo polymerization generally needs to be conjugated to other groups (i.e. styrene, methyl methacrylate, etc.) such that the monomer readily absorbs visible or ultra-violet light [60]. In the case of poorly absorbing monomers, a photosensitizer can be used which absorbs the light and then activates the monomeric unit to be polymerized [61]. Absorption of light by either the sensitizer or monomer thus generates an excited state that either decomposes into radical groups or undergoes reaction with a second species to generate radical species [60, 61]. The corresponding radical species then become the formal initiators for polymerization. The photopolymerization of styrene dates back to 1839 [62, 63].

In the case of thermal polymerization, heating generally results in the breaking of a bond which generates radical species [64], although the generation of radicals via other types of thermal reactions have also be reported [65]. These radical species then serve as the polymerization initiators. As such, thermal initiation is typically limited to compounds with low bond dissociation energies ($100–170$ kJ mol^{-1}) [64], which generally requires an initiating species containing a O–O, S–S, or N–O bond. Initiation of compounds with higher bond dissociation energies are still possible, but typically undergo dissociation too slowly or require unreasonably high temperatures to induce polymerization.

The term *polyethylene* was first introduced by Berthelot in 1869 [66]. It had long been recognized that during the production of ethylene from alcohol and sulfuric acid, a liquid fraction was also produced that was referred to as *oil of wine*. Oil of wine consisted primarily of various higher olefins, which Berthelot isolated and fractionated in 1869 in order to identify the various olefin species contained in the oil. He determined that the highest boiling fraction was hexadecene ($C_{16}H_{32}$ or $[C_2H_4]_8$), which he then referred to as polyethylene. Although the product here was still of relatively low molecular weight, this example could be described as an example of cationic polymerization.

Solid polyethylene was first prepared in 1898, although not via the polymerization of ethylene. As part of his investigations of diazomethane (CH_2N_2), Hans von Pechmann (1850–1902) observed the formation of small quantities of a white, flakey substance from a standing ether solution of diazomethane [67]. Shortly thereafter, Eugen Bamberger (1857–1932) and Friedrich Tschirner isolated the same material as a byproduct during reactions utilizing diazomethane and reported its characterization in 1900 [68]. They found the solid to be insoluble in ordinary organic solvents, with solubility only observed in either boiling cumene or boiling pyridine. After purification via precipitation from cumene by petroleum ether addition, they dried the material, which was described as an amorphous, chalk-like powder. The solid was found to melt at 128 °C and analysis gave an empirical formula of CH. From these results, they gave it the formula $(CH_2)_x$ and named in *polymethylen* [68].

The production of solid polyethylene via the polymerization of ethylene wasn't until 1933, when high pressure reactions were being investigated at Imperial Chemical Industries (ICI). An attempt to react ethylene with benzaldehyde at high temperature and pressure produced a waxy solid that was identified as polyethylene [69]. This process was then repeated in 1935, when polyethylene was produced from pure ethylene at 1400 atm and 170 °C. After further optimization, a patent was granted to ICI in 1939 [70].

1.4 Scope of the Current Volume

The history of acetylene polymers is dominated by the semiconducting material polyacetylene, which became the basis for the 2000 Nobel Prize in Chemistry. Awarded to Professors Alan J. Heeger, Alan G. MacDiarmid, and Hideki Shirakawa "for the discovery and development of electrically conductive polymers," the Nobel Prize was in acknowledgement of their early contributions to the field of conjugated organic polymers, particularly their collaborative work on conducting polyacetylene beginning in the mid-to-late 1970s [69, 71, 72]. As a result, the history of polyacetylene in the 1970s, particularly the discovery of metallic conductivity via doping, has been documented and reported by a number of authors [69, 71–78]. However, very little has been documented in terms of the history of polyacetylene prior to the work of Shirakawa [71]. Of course, even less attention has been given to the history of acetylene polymerizations prior to the work of Natta in 1955, with only a few papers on the history of the material that came to be known as cuprene [79, 80]. As such, the goal of the current volume is to provide the first comprehensive history of acetylene polymers, beginning with the first polymerization studies by Berthelot in 1866 and continuing up through the 2000 Nobel Prize in Chemistry. In addition to covering the history of formal acetylene polymers, the current volume will also present the history of two related materials, polyenes and polyvinylenes, which are structurally analogous to polyacetylene, but are not produced through the polymerization of acetylene. In the process, the reader will hopefully gain insight into how the histories of these various materials are interrelated, as well as the realization that acetylene polymers have a much longer and richer history than commonly believed.

References

1. Crosland M (2000) "Slippery Substances": some practical and conceptual problems in the understanding of gases in the pre-lavoisier era. In: Holmes FL, Levere TH (eds) Instruments and experimentation in the history of chemistry. The MIT Press, Cambridge, Massachusetts, pp 79–104
2. Rosenfeld L (1985) The last alchemist—The first biochemist: J. B. van Helmont (1577–1644). Clin Chem 31:1755–1760
3. Partington JR (1989) A short history of chemistry, 3rd edn. Dover, New York, pp 44–49
4. Stillman JM (1924) The story of early chemistry. D. Appleton and Co., New York, pp 381–386
5. Brock WH (1992) The Norton history of chemistry. W. W. Norton & Company, New York, pp 49–52
6. Jensen WB (2004) A previously unrecognised portrait of Joan Baptista Van Helmont (1579–1644). Ambix 51:263–268
7. Burget GE (1925) Stephen Hales (1677–1761). Ann Med Hist 7:9–20
8. Clark-Kennedy AE (1977) Br Med J 2:1656–1658
9. Partington JR (1989) A short history of chemistry, 3rd edn. Dover, New York, pp 90–93
10. Brock WH (1992) The Norton history of chemistry. W. W. Norton & Company, New York, pp 77–78
11. Partington JR (1998) A history of chemistry. Martino Publishing, Mansfield Centre, CT, vol 4, pp 432–433
12. Christoph Meinel C (1992) August Wilhelm Hofmann-"Reigning Chemist-in-Chief'. Angew Chem Int Ed 31:1265–1398
13. Hofmann AW (1866) On the action of trichloride of phosphorus on the salts of the aromatic monamines. Proc R Soc London 15:54–62
14. Verkade PE (1985) A history of the nomenclature of organic chemistry. D. Reidel Publishing Company, Dordrecht, p 23
15. Verkade PE (1985) A history of the nomenclature of organic chemistry. D. Reidel Publishing Company, Dordrecht, p 174
16. Partington JR (1998) A history of chemistry. Martino Publishing, Mansfield Centre, CT, vol 4, p 61
17. Trasatti S (1999) 1799–1999: Alessandro Volta's 'Electric Pile' two hundred years, but it doesn't seem like it. J Electroanal Chem 460:1–4
18. Partington JR (1998) A history of chemistry. Martino Publishing, Mansfield Centre, CT, vol 4, p 6
19. Volta A (1777) Sull' Aria Inflammabile Nativa delle Paludi. Giuseppe Marelli, Milan
20. Lappert MF, Murrell JN (2003) John Dalton, the man and his legacy: the bicentenary of his Atomic Theory. Dalton Trans 2003:3811–3820

21. Rocke AJ (2013) The Quaker Rustic as natural philosopher: John Dalton and His Social Context. In: Patterson GD, Rasmussen SC (eds) Characters in chemistry: a celebration of the humanity of chemistry. In: ACS symposium series 1136, American Chemical Society, Washington, DC, pp 49–60
22. Dalton J (1808) New system of chemical philosophy. R. Bickerstaff, Manchester, vol 1, pp 444–450
23. Berthelot M (1856) Synthèse des carbures d'hydrogène. C R Hebd Seances Acad Sci 43:236–238
24. Partington JR (1998) A history of chemistry. Martino Publishing, Mansfield Centre, CT 4:468
25. Meadows J (1992) The great scientists. The story of science told through the lives of twelve landmark figures. Oxford University Press, New York, pp 129–148
26. Jensen WB (1991) Michael Faraday and the art and science of chemical manipulation. Bull Hist Chem 11:65–76
27. Faraday M (1834) Experimental researches in electricity—Seventh Series. Phil Trans R Soc Lond 124:77–122
28. Partington, JR (1998) A history of chemistry. Martino Publishing, Mansfield Centre, CT, vol 4, pp 500–502
29. Pilcher RB (1914) History of the institute: 1877–1914. The Institute of Chemistry of Great Britain and Ireland, London, p 50
30. Partington JR (1998) A history of chemistry. Martino Publishing, Mansfield Centre, CT, vol 4, pp 502–503
31. Frankland E, Kolbe H (1848) Ueber die Zersetzungsproducte des Cyanäthyls durch Einwirkung von Kalium. Ann Chem Pharm 65:269–287
32. Frankland E, Kolbe H (1849) On the products of the action of potassium on cyanide of ethyl. Quart J Chem Soc London 1:60–74
33. Frankland E (1849) Ueber die Isolirung der organischen Radicale. Ann Chem Pharm 71:171–213
34. Frankland E (1850) On the Isolation of the Organic Radicals. Quart J Chem Soc London 2:263–296
35. Frankland E (1951) Researches on the organic radicals. Quart J Chem Soc London 3:322–347
36. Partington JR (1998) A history of chemistry. Martino Publishing, Mansfield Centre, CT, vol 4, pp 507
37. Berthelot M (1857) Substitutions inverses. Ann Chim 51:48–58
38. Berthelot M (1867) Nouvelles applications des méthodes de réduction en chimie organique. Bull Soc Chim 7:53–65
39. Roscoe HE, Schorlemmer, C (1878) A treatise on chemistry. D. Appleton and Company, New York, vol 1, pp 611–613
40. Gest H (2000) Bicentenary homage to Dr. Jan Ingen-Housz, MD (1730–1799), pioneer of photosynthesis research. Photosynth Res 63:183–190
41. Ingen Housz JM, Beale N, Beale E (2005) The life of Dr. Jan Ingen Housz (1730–99), private counsellor and personal physician to Emperor Joseph II of Austria. J Med Biogr 13:15–21
42. Ingen-Housz J (1779) Account of a new kind of inflammable Air or Gas, which can be made in a moment without apparatus, and is as fit for explosion as other inflammable gasses in use for that purpose; together with a new theory of gun powder. Philos Trans R Soc Lond 69:376–418
43. Partington JR (1998) A history of chemistry. Martino Publishing, Mansfield Centre, CT, vol 3, pp 584–585
44. Deiman JR, Van Troostwyck AP, Bondt N, Lauwerenburgh A (1794) Sur les diverses espèces des Gaz qu'on obtient en mêlant l'acide sulfurique concentré avec l'alkool. J Phys Chim 45:178–191
45. Davy E (1836) Notice of carburet of potassium, and of a new gaseous bi-carburet of hydrogen. Rec Gen Sci 4:321–323
46. Davy E (1837) Notice of a peculiar compound of carbon and potassium, or carburet of potassium, & c. Br Assoc Adv Sci Rep 5:63–64

47. Berthelot M (1860) Sur une nouvelle série de composés organiques, le quadricarbure d'hydrogène et ses dérivés. C R Acad Sci Paris 50:805–808
48. Partington JR (1998) A history of chemistry. Martino Publishing, Mansfield Centre, CT, vol 4, p 476
49. Reynolds JW (1951) On "Propylene," a new hydrocarbon of the series C_nH_n. Quart J Chem Soc Lond 3:111–120
50. Johnstone GA (1927) Advantages of ethylene-oxygen as a general anesthetic. Cal West Med 27:216–218
51. Morris PJT (1986) Polymer pioneers. Center for the History of Chemistry, Phiadelphia, pp 48–50
52. Morawetz H (1985) Polymers. The Origins and Growth of a Science. Wiley, New York, pp 86–87
53. Dorel Feldman D (2008) Polymer history. Des Monomers Polym 11:1–15
54. Nicholson JW (1991) Etymology of 'polymers'. Educ Chem 28:70–71
55. Jensen WB (2008) The origin of the polymer concept. J Chem Ed 88:624–625
56. Partington JR (1998) A history of chemistry. Martino Publishing, Mansfield Centre, CT, vol 4, pp 142–144
57. Berzelius JJ (1933) Isomerie, Unterscheidung von damit analogen Verhältnissen. Jahr Ber For Phys Wiss 12:63–67
58. Odian G (2004) Principles of polymerization, 4th edn. Wiley-Interscience, Hoboken, NJ, pp 4–7
59. Chanda M (2013) Introduction to polymer science and chemistry, 2nd edn. CRC Press, Boca Raton, FL, pp 7–10
60. Odian G (2004) Principles of polymerization, 4th edn. Wiley-Interscience, Hoboken, NJ, pp 218–220
61. Chanda M (2013) Introduction to polymer science and chemistry, 2nd edn. CRC Press, Boca Raton, FL, pp 308–309
62. Simon S (1839) Ueber den flüssigen Storax (Styrax liquidas). Ann Pharm 31:265–277
63. Blyth S, Hofmann AW (1845) Ueber das Styrol und einige seiner Zersetzungsproducte. Ann Chem Pharm 53:289–329
64. Odian G (2004) Principles of polymerization, 4th edn. Wiley-Interscience, Hoboken, NJ, pp 209–211
65. Chanda M (2013) Introduction to polymer science and chemistry, 2nd edn. CRC Press, Boca Raton, FL, pp 312–313
66. Berthelot M (1869) Méthode universelle pour réduire et saturer d'hydrogène les composés organiques. Bull Soc Chim France 11:4–35
67. von Pechmann H (1898) Ueber Diazomethan und Nitrosoacylamine. Ber Dtsch Chem Ges 31:2640–2646
68. Bamberger E, Tschirner F (1900) Ueber die Einwirkung von Diazometban auf β-Arylhydroxylamine. Ber Dtsch Chem Ges 33:955–959
69. Rasmussen SC (2011) Electrically conducting plastics: revising the history of conjugated organic polymers. In: Strom ET, Rasmussen SC (eds) 100+ years of plastics: Leo Baekeland and Beyond, ACS symposium series 1080. American Chemical Society, Washington, DC, pp 147–163
70. Morawetz H (1985) Polymers. The Origins and Growth of a Science. Wiley, USA, p 132
71. Rasmussen SC (2014) The path to conductive polyacetylene. Bull Hist Chem 39:64–72
72. Rasmussen SC (2017) Early History of Conductive Organic Polymers. In: Zhang Z, Rouabhia M, Moulton SE (eds) Conductive polymers: electrical interactions in cell biology and medicine. CRC Press, Boca Raton, FL, 2017; Chapter 1
73. Shirakawa H (2001) The discovery of polyacetylene film: the dawning of an era of conducting polymers (Nobel Lecture). Angew Chem Int Ed 40:2574–2580
74. Shirakawa H (2001) The discovery of polyacetylene film: the dawning of an era of conducting polymers. Synth Met 125:3–10
75. Hall N (2003) Twenty-five years of conducting polymers. Chem Commun 2003:1–4

76. Hargittai I (2011) Drive and curiosity: what fuels the passion for science. Prometheus Books, Amherst, NY, pp 173–190
77. Elschner A, Kirchmeyer S, Lovenich W, Merker U, Reuter K (2011) PEDOT: principles and applications of an intrinsically conductive polymer. CRC Press, Boca Raton, FL, pp 1–20
78. Heeger AJ (2016) Never lose your nerve! World Scientific Publishing Co., Singapore, pp 131–150
79. Calhoun JM (1937) A study of cuprene formation. Can J Res 15b:208–223
80. Rasmussen SC (2017) Cuprene: a historical curiosity along the path to polyacetylene. Bull Hist Chem 42:63–78

Chapter 2
Acetylene

Pure acetylene is a colorless gas that is said to have a slight, pleasant odor. The characteristic disagreeable odor of common acetylene is due to trace impurities resulting from its preparation from calcium carbide. Physiologically, acetylene acts as an anesthetic and purified acetylene gas has been used for this function [1]. Acetylene can be produced via a number of methods including the hydrolysis of alkali and alkaline-earth metal carbides, the elimination of halides and haloacids from various organohalides, and through the direct synthesis from elemental carbon and hydrogen. The earliest known production of acetylene by any method was reported by Edmund Davy (1785–1857) in 1836 [2–7].

2.1 Edmund Davy and Bicarburet of Hydrogen

Edmund Davy was born in 1785 at Penzance, in Cornwall [7, 8]. His father was William Davy and his uncle was Robert Davy. As such, Edmund was the cousin of the more well-known chemists Dr. John Davy (1790–1868) and Sir Humphry Davy (1778–1829) [7–10]. Edmund obtained his early education in Penzance, before moving to London to join his cousin Humphry at the Royal Institution [7, 9]. After dismissing an assistant for "idleness and general neglect of duty," Humphry recommended Edmund to the managers of the Royal Institution on January 12, 1807 as a "young man of good conduct and some promise" [8]. Thus, Edmund was given a room on the Attic Floor and he became an assistant under Humphry with the duty of looking after the laboratory. Six months later, Edmund was given a better room and, in January 1809, was given a raise in pay due to additional duties consisting of looking after and showing the Royal Institution's mineralogical collection [7, 8]. Although Humphry had high standards concerning the state of the laboratory, Edmund managed to meet them and he remained at the Royal Institution with additional raises in pay until the end of 1812 [7–9]. It has been said that Edmund resigned due to

© The Author(s) 2018
S. C. Rasmussen, *Acetylene and Its Polymers*, SpringerBriefs in Molecular Science,
https://doi.org/10.1007/978-3-319-95489-9_2

dissatisfaction with changes following Humphry's marriage and departure from the Royal Institution [8].

Edmund then moved to Cork, in Ireland, were he was unanimously elected professor at the Royal Cork Institution in 1813 [7, 8]. It has been proposed that this may have been through the influence of his cousin Humphry, who had delivered a course of lectures in Dublin two years previously [7]. He then moved to Dublin in 1826 to become professor of chemistry in the Royal Dublin Society. Shortly thereafter he was elected a fellow of the Royal Society of London on January 19, 1826. He was also elected a fellow of the Chemical Society of London and an honorary member of the Société Française Statistique Universelle [7]. Edmund remained in Dublin until his retirement.

Edmund had a successful career and published 33 papers over the period of 1812 to 1857. Upon his retirement, the government awarded him his entire salary in recognition of his accomplishments. After June of 1856, Edmund suffered from ill-health until his death the following year at Kimmage in County Dublin on November 5, 1857 [7].

On August 26, 1836, at the Sixth Meeting of the British Association for the Advancement of Science in Bristol, Davy presented results of a new gaseous compound of carbon and hydrogen [2]. The details of this report were then published in Thomson's *Records of General Science* [3] in late 1836, followed quickly by French and German versions published in *Journal de Pharmacie et de Chimie* [4] and *Annalen der Pharmacie* [5] in 1837.

These reports presented the results of a study that began with efforts to produce potassium metal on a large scale in January of 1836. Such efforts involved the high temperature heating of a mixture of tartar (potassium hydrogen tartrate, $KC_4H_5O_6$) and charcoal powder in an iron bottle [2–5]. In this way, Davy obtained a dark-gray substance that was described as having a granular structure, but a rather soft form that could be easily cut with a knife. When added to water, this substance vigorously decomposed to generate carbonaceous matter with the evolution of large amounts of gas, along with occasional inflammations on the water surface. Analysis of the evolved gas led to the conclusion that it was a nearly equal volume mixture of hydrogen and "a new bi-carburet of hydrogen" [3], the latter consisting of carbon and hydrogen in nearly equal volumes [2]. From these results, Davy viewed the original gray substance as a mixture of potassium and carburet of potassium [2–5].

In one experiment, heating the tartar-charcoal mixture led to no potassium, but the small quantity of a dark black substance which Davy believed to be solely the carburet of potassium produced in the previous example. To the naked eye, this solid did not appear crystalline, but minute, truncated four-sided prisms could be seen under strong magnification [2–5]. When this solid was added to water, the new bicarburet of hydrogen was again produced, this time as the only gaseous product. Davy then concluded that pure carburet of potassium was a binary compound consisting of one proportion of carbon and one of potassium (what is now known as potassium carbide or potassium acetylide, K_2C_2) [2–5].

Davy then went on to investigate the new gaseous bicarburet of hydrogen obtained by the action of water on carburet of potassium. The gas was highly flammable and, in the presence of air [3–5], it burned with a bright flame described to be "denser and of greater splendour than even olefiant gas" [3].[1] This gas reacted explosively with chlorine gas, producing a large red flame with the deposition of carbon. This reaction with chlorine was found to take place even in the dark and was thus independent of the action of light [3–5]. Davy's bicarburet of hydrogen could be stored over mercury for an indefinite time without reaction, but if stored over water, the gas was slowly absorbed up to about an equal volume of the water. Upon heating the aqueous solution, the gas evolved without apparent reaction [3–5].

The gas detonated in the presence of oxygen to give water and carbonic acid. Its complete combustion required 2.5 volumes of oxygen, two volumes of which are converted into carbonic acid, and the remaining half volume into water [3–5]. From these various analyses, Davy concluded that the new gas was composed of one volume of hydrogen and two volumes of carbon [3–5], stating [3].

> It is, in fact, a bi-carburet of hydrogen composed of two proportions of carbon and one of hydrogen, and may be represented by the formula $C^2 + H$, or $2C + H$; and its constitution seems to differ from that of any other known gas.

Lastly, Davy proposed the application of the new gas for gaslight [3]:

> From the brilliancy with which the new gas burns in contact with the atmosphere, the author thinks it is admirably adapted for producing artificial light, if it can be procured at a cheap rate.

On June 26, 1837, Davy then gave a second presentation, communicating a paper entitled "On a new Gaseous Compound of Carbon and Hydrogen" at a Scientific Meeting of the Royal Dublin Society [11]. He also gave a very brief update of his investigations on the new gas at the Seventh Meeting of the British Association for the Advancement of Science during September of 1837 in Liverpool [12]. These presentations were then followed by the publication of his final and most detailed paper on the topic in 1839 [13].

The initial publication from the *Proceedings of the Royal Irish Academy* did not offer much new data and primarily repeated material reported in the previous publications. However, it did provide a value of 0.917 for the specific gravity of the gas, in comparison to 1.000 for air as a standard.[2] In addition, it formalized Davy's proposed name for the gas as *bicarburet of hydrogen* [11].

The report from the British Association meeting was very brief and only reported results from passing electrical sparks through the gas, which resulted in the deposition of carbon with no change in gas volume [12]. Although he later stated that he had initially thought the gas volume after the reaction to be hydrogen [13], he ultimately decided that it was another new hydrocarbon gas. Unlike his bicarburet of hydrogen, this new gas did not ignite in contact with chlorine and required only 1.5 volumes of oxygen for complete combustion, again producing only carbonic acid and water

---

[1] The species known as olefiant gas is the modern-day ethylene or ethene.

[2] The modern density value for acetylene is 1.097×10^{-3} g/mL, while the density of air is 1.2041×10^{-3} g/mL at 20 °C. Thus, using air as a standard, this would give a specific gravity of 0.911 for acetylene, in very good agreement with Davy's value.

vapor. Davy believed this gas to be a binary compound represented by the formula C+H [12].

In the final paper in the *Transactions of the Royal Irish Academy*, Davy provided detailed procedures for generating and isolating the gas from the previously described carburet of potassium. Using these methods, Davy found that six grains of the solid produced about two cubic inches of the gas [13]. He then continued in describing the properties of the gas, beginning with a comparison of its combustion to that of olefiant gas. While olefiant gas burned with a bluish flame, the new gas burned with a bright white flame. Additional new data reported included the fact that aqueous solutions of the gas had no smell or taste, and caused no effect to litmus paper [13]. The remainder of the characterization data had been previously reported in his earlier papers. Davy then concluded with detailed experiments to determine the composition of his bicarburet of hydrogen, concluding in the same formula as he had previously reported (2C+H). However, he correctly determined that his gas had one less proportion of hydrogen than olefiant gas, which he viewed as 2C+2H (rather than the modern formula C_2H_4) [13].

It should be pointed out that during the time period under discussion, molecular weights and thus molecular formulas were still a developing aspect of the chemical sciences and neither accurate weights nor formulas had yet been fully developed [14]. A common example was that many still viewed the molecular formula of water as the simplest combination of hydrogen and oxygen, that is HO, rather than H_2O. As a result, the weight of oxygen relative to hydrogen was then viewed as 8, rather than 16, and application of an oxygen value of 8 would then give a carbon value of only 6. It was these values that were commonly used in England until the 1860s [14] and the use of a weight of 6 for carbon would usually result in a doubling of the number of carbons in the corresponding molecular formulas, thus accounting for the reported formulas of Davy.

Although Davy provided significant characterization data for this new gas, he was not able to establish its structure, nor did his name for the gas endure. In addition, although he had suggested its use for lighting, it was not applied as such until the 1890s. As a result, Davy's gas was essentially forgotten until Marcelin Berthelot (1827–1907) rediscovered it in 1860 [15].

2.2 Marcelin Berthelot and Acetylene

Pierre Eugène Marcelin Berthelot[3] (Fig. 2.1) was born in Paris to parents Jacques Martin Berthelot (1799–1864) and Ernestine Sophie Claudine Berthelot (née Biard) (1800–1876) on October 25, 1827 [16–19]. His father was a medical doctor [16, 17] and the family lived in the heart of Paris at the Place de Grève (now the Place de

---

[3]His name appears as both 'Marcelin' or 'Marcellin' in the literature. Berthelot signed 'Marcellin' on his first publication in 1850 and he used this spelling in most of his scientific writings, but he signed authorized documents as 'Marcelin Berthelot' [16].

Fig. 2.1 Pierre Eugène
Marcelin Berthelot
(1827–1907) (Edgar Fahs
Smith Memorial Collection.
Kislak Center for Special
Collections, Rare Books and
Manuscripts. University of
Pennsylvania)

l'Hôtel de Ville) [17]. After attending a neighborhood elementary school, Marcelin entered the Collège Henri IV (renamed Lycée Henri IV as of 1873) in 1843 [16–18]. He received the *Baccalauréat ès Lettres* required for entrance to the university in 1846, after which was also awarded the *Baccalauréat ès Sciences* in 1848 [16]. He then continued his studies at the Collège de France [16–19]. As the holder of two *Baccalauréats*, Marcelin was entitled to undertake studies in both the arts and science, but he ultimately focused on science [16]. To satisfy his father, Marcelin registered at the Faculté de Médecine in November of 1848, but he resigned in 1850. He had also enrolled simultaneously at the Faculté des Sciences, however, and he earned the degree of *Licence ès Physique* on July 26, 1849 [16].

Marcelin then entered a private school for the practical teaching of chemistry founded on the Rue Dauphine by Théophyle Jules Pelouze (1807–1867), a lecturer at the Collège de France and former assistant to Joseph Louis Gay-Lussac (1778–1850). At the school, Marcelin supervised the students' work, but was otherwise free to experiment and he published his first two articles in 1850 [16, 17]. Marcelin only spent a short time at Pelouze's school, however, as Pelouze resigned from the Collège in 1851 to take another position [16].

Due to a recommendation from Pelouze to Antoine Balard (1802–1876), his replacement at the Collège de France, Marcelin then returned to the Collège in February 1851 as Balard's assistant in charge of lecture demonstrations [16, 17]. Three years later, in April of 1854, he submitted his thesis for the *Doctorat ès Sciences* [16, 17], after which he registered at the Ecole Supérieure de Pharmacie. In November of 1858, he submitted a second thesis for the *Doctorat en Pharmacie*, while also acquiring the diploma of *Pharmacien de Première Classe* [16]. Due to the intervention of Jean-Baptiste Dumas (1800–1884), professor at the Collège de France and General Inspector of Public Education, a chair of organic chemistry was created at the Ecole Supérieure de Pharmacie in December of 1859. Coincidentally, Marcelin had just acquired the prerequisites for this position and he thus became the first titular professor of the chair and held the position until 1876 [16, 19]. A laboratory associated with the new chair was not available for a number of years, however, so he pursued his research at the Collège de France [16].

Due to interventions by Dumas and Balard, Marcelin was then entrusted with the lectures on organic chemistry at the Collège de France in 1863. This was in addition to his normal teaching duties at the Ecole Supérieure de Pharmacie [16]. A new chair of organic chemistry was then created for him at the Collège in August of 1865 [16, 18]. He retained this chair until his death on March 18, 1907 [16, 18, 19].

It was in 1860 [15], shortly after assuming the chair of organic chemistry at the Ecole Supérieure, that Berthelot reported studies on what he believed to be a new hydrocarbon gas which he gave the name *acétylène*[4] [15, 22]. It should be pointed out that as of his 1860 publication, Berthelot was not aware of Davy's earlier discovery. By 1863, however, he did recognize Davy's previous discovery in a review of this work on acetylene, stating [22]:

> Edm. Davy obtained this gas in 1836 by treating with water the black mass which occurs in the preparation of potassium by means of calcium tartrate and charcoal. But his observation, which had remained isolated, had disappeared from science: I was not aware of it when I found the same gas by very different methods.

Although Davy's work on the gas had been published extensively over the period of 1836–1839, this was not followed up with any further studies and thus his work had faded from memory as pointed out by Berthelot. Berthelot's rediscovery of the gas in 1860 marked the beginning of the significant study and application of acetylene.

---

[4]Although Berthelot did not explain the reasoning for the name, it is generally believed to be the combination of *acetyl* with the suffix *-ene*. The acetyl radical, C_4H_6 (the modern C_2H_3), was first proposed by Justus von Liebig (1803–1873) in 1839 [20]. The name of this radical was derived from acetic acid as the acid was considered the oxide of the acetyl radical [21]. In a similar manner, if one hydrogen were added to acetyl, this would give ethylene and if two were added, this would give the ethyl radical [20, 21]. Thus, acetylene was presumably named in analogy to ethylene, whereas ethylene was the ethyl radical less one hydrogen, and acetylene was the acetyl radical less one hydrogen. At this point in time, the suffix *-ene* did not have any specific meaning and did not specifically refer to double-bonded species until after 1866.

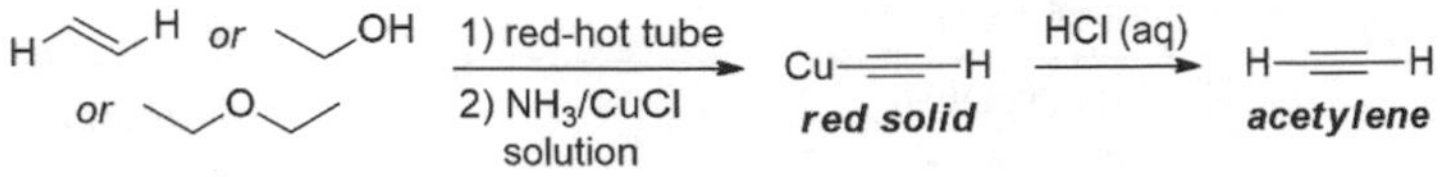

Fig. 2.2 Berthelot's initial synthesis of acetylene

Berthelot was able to produce the gas by passing various organic gases or vapors (ethylene, alcohol, ether, aldehyde, etc.) through a red-hot tube. The initial gas produced was found to be a complex mixture, thus requiring him to trap the acetylene as an intermediate species by passing the gaseous mixture into an ammonia solution of cuprous chloride (Fig. 2.2). This resulted in the precipitation of a red copper acetylide, which could then be collected and purified. When this red solid was treated with hydrochloric acid, this would then liberate the acetylene gas [15, 19, 22]. The final gas product was then purified by washing with a little potash [15, 22].

Berthelot described the collected gas as colorless, sparingly soluble in water, and endowed with a characteristic and unpleasant odor. He found that it burned with a very bright and smoky flame [15, 22]. As previously described by Davy [3–5], he found that when mixed with chlorine, the gas detonated almost immediately with the deposition of charcoal, even in the absence of direct light [15]. He was not able to liquefy the gas by either cold or pressure, but determined its relative density (specific gravity) to be 0.92 [15].

He continued his study by characterizing the combustion of the gas, and as previously determined by Davy [3–5], found that the complete combustion of one volume of acetylene required 2.5 volumes of oxygen. The product of this combustion was found to be two volumes of carbonic acid. This data, along with the previously determined density, led Berthelot to the conclusion that acetylene was the species hydrogen tetracarbide, represented by the formula C_4H_2 [15].

Lastly, he investigated its chemical reactivity and the production of various derivatives. Overall, he found that acetylene possessed most of the essential properties of ethylene and furnished paralleled derivatives when reacted with bromine and sulfuric acid. To further strengthen the relationship between acetylene and ethylene, Berthelot then showed that treatment of acetylene with hydrogen at low temperature resulted in the production of ethylene [15, 22]. He did not study these reactions or their products in great detail, however, stating that difficulties in preparing large quantities of acetylene prevented more extensive investigations.

Berthelot then followed up this initial report with a second paper in early March of 1862 [23], which attempted to place acetylene in the context of other known carbon-hydrogen species of the time (methane, ethylene, propylene, etc.). This discussion focused on various synthetic conversions, showing that methane could be used to generate acetylene via either heat or spark. The generated acetylene could then be converted to ethylene by reaction with hydrogen [22, 23]. This last was accomplished by treating the copper acetylide intermediate with zinc in aqueous ammonia, thus generating a mixture of ethylene and hydrogen (Fig. 2.3).

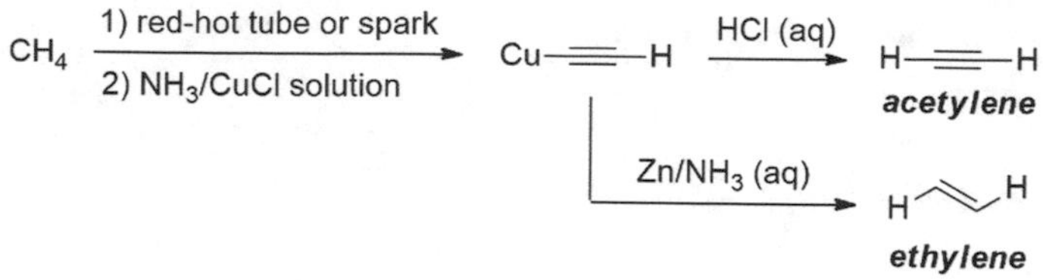

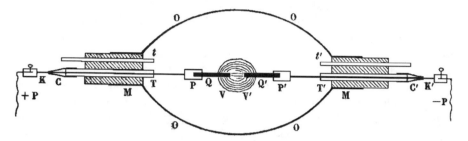

Fig. 2.3 Interconversion of various hydrocarbons

Fig. 2.4 Marcelin's *d'œuf électrique* (electric egg) (From Berthelot [26])

A few weeks later, in late March of 1862 [16, 24], Berthelot then reported the production of acetylene via an electric discharge between two carbon rods in the presence of hydrogen [16, 22, 24–26]. These activities utilized what was often referred to as Marcelin's *d'œuf électrique* (electric egg), shown in Fig. 2.4 [16, 26]. This ellipsoid-shaped device is equipped with openings at the ends of the long axis, which are sealed with fat plugs (labeled M in Fig. 2.4). Each of these plugs is fitted with two glass tubes, one of which (*t* and *t′*) allows one to feed hydrogen gas into the device. The other tube (T and T′) is fitted with a metal rod such that it can slide within the glass tube with gentle friction, while still providing a seal between the interior of the vessel and the exterior atmosphere. The outer end of the metal rod is connected to a pole of a battery, while the inner end is fitted with a connector (P and P′) capable of holding a carbon rod (Q and Q′) [26]. Berthelot stressed the importance of purifying the carbon to be utilized in the rods by heating the carbon in a porcelain tube under a stream of dry chlorine until it becomes red-white [22, 24–26].

Under operation, the device is first filled with hydrogen and metal rods are connected to the appropriate poles of the battery. The carbon rods are then brought together until they touch, after which they are slightly displaced in order to allow the electric arc between them. As the process continues, the tips of the carbon rods are consumed and care must be taken to adjust the distance between the carbon rods to reestablish the arc [26]. Approximately half the carbon is converted into acetylene, while the other half is dispersed into dust which adherents to the interior walls of the device.

As before, the acetylene product was trapped by passing the gaseous volume of the device into an ammonia solution of cuprous chloride to generate the red copper acetylide intermediate [22, 24–26]. The isolated and purified copper acetylide was then treated with HCl to liberate the pure acetylene gas. This now allowed Berthelot to generate significant amounts of acetylene, with a reported production rate of 10–12 mL per min [22, 24–26].

The priority of this discovery, however, was questioned by Dean Morren of the Faculté des Sciences in Marseille [16], who had presented a brief report in 1859 on the production of gaseous species under the influence of electricity. Included in this report was a single line pertaining to a gaseous hydrocarbon [27]:

> By taking carbon electrodes and circulating hydrogen, I obtained a carbonaceous hydrogen of which I have not yet verified the special nature.

Thus, following Berthelot's 1862 report [24], Morren presented a case that he had now confirmed that the gas he had obtained in 1859 was positively acetylene [28]. Berthelot quickly responded, stating that Morren's new report was quite different from that of 1859 and seemed influenced by the details of his own recent report [29]. He then put an end to Morren's claim of priority with the statement [29]:

> It is not that I pretend to question Mr. Morren's good faith. But whatever may have been his experience of 1859, the means of purification (the elimination of hydrogen from coal by free hydrogen) and the analytical methods which he has since announced were absolutely insufficient to establish the production of a carbonaceous hydrogen.

Even this statement, however, seems to be giving Morren too much credit as all he ever reported in 1859 was a single claim with little detail of the experimental conditions and absolutely no evidence of the nature of the product, if any, produced.

Following his March 1862 report of the synthesis of acetylene via electric arc, Bertholet then presented three additional papers in May of the same year [30–32]. The first of these reports focused on the effectiveness of different types of carbon sources when used for the carbon electrodes in the electrolytic synthesis of acetylene [30]. The second reviewed all of the successful methods for the production of acetylene, as well as those methods that had failed to generate acetylene [31]. The final paper reported confirmation that lighting gas[5] contained acetylene as a component, although its relative proportion in the gas mixture was fairly low [32]. However, the small amount of acetylene in the mixture was believed to provide much of the resulting flame's considerable illumination, as well as the distinctive smell of lighting gas.

Although the electric discharge process of Berthelot now allowed the production of greater amounts of acetylene than previously possible, it was still not a suitable method for the mass production of the gas. Such large-scale manufacture of the gas would have to wait until the 1890s, when Thomas L. Wilson (1860–1915) developed a process for the large-scale production of calcium carbide from lime and coke.

---

[5] Although it is not specified, it is thought that Berthelot is referring specifically to coal gas, which was the oldest and most cost-effective form of gas for lighting at the time.

Fig. 2.5 Thomas Leopold Willson (1860–1915) (Library and Archives Canada/C-53499)

2.3 Thomas Willson and Acetylene from Calcium Carbide

Thomas Leopold Willson (Fig. 2.5) was born March 14, 1860 near Princeton, in the region of Canada that is now Ontario [33–36]. His parents Thomas Whitehead Willson and Rachel Sabina Bigelow ran a farm near Princeton [34] and his grandfather was John Willson, Speaker of the United Canadian Assembly [33, 34]. A bad investment led to the loss of the farm and the family moved to Bridgeport, Connecticut, before returning to Hamilton, Ontario around 1872 [34]. Willson entered Hamilton Collegiate Institute in 1876 [33–36], but withdrew from school after his father died in 1879 [33, 35]. He was then apprenticed to John Rodgers, a blacksmith who allowed him to develop his own inventions in the loft of the smithy [34, 36]. One such invention was an arc-lighting system, the first seen in Hamilton [33–35].

At age 22, Willson then moved to the United States, [33–36] where he held various jobs in the mechanical and electrical trades before settling in Brooklyn, New York, in 1887 [33, 35]. At the same time, he continued to work on his own projects, resulting in six patents over the next three years and securing the US rights for use of the

Fig. 2.6 James Turner Morehead (1840–1908) (From Samuel A. Ashe SA, ed (1905) Biographical History of North Carolina. Charles L. Van Noppen, Greensboro, N.C.)

electric-arc furnace in ore smelting [33, 35]. Various efforts to produce his designs through other parties never resulted in a marketable product, so he formed his own company, Willson Electric, in 1890. It too was unsuccessful, though, in part because manufacturers were nervous about investing in untried electric technologies [36].

In December 1890, Willson formed the Willson Aluminum Company in partnership with James Turner Morehead (1840–1908) (Fig. 2.6) [33–36]. Morehead had surplus waterpower at his cotton mill in Spray, North Carolina, and thus Willson moved to Spray in the autumn of 1891 to build a small 300-horsepower plant along the Smith River on Morehead's land [33–36]. The goal of the new company was to develop an inexpensive means for the production of pure aluminum. One approach apparently tried by Willson was the use of calcium to reduce aluminum chloride in an electric furnace. In the process, he accidently produced calcium carbide, which produced acetylene when added to water [32–40].

Towards the end of 1894, Willson sold his American patents and then returned to Canada. In Canada, Willson formed the Willson Carbide Works Company of St. Catharines, Ontario, which built its first plant in 1895 at nearby Merritton, where waterpower from the Welland Canal was used to generate the needed electricity [33, 34, 36]. Willson later had more plants constructed in Ottawa and Shawinigan Falls, Quebec [34–36].

After moving to Ottawa in 1901 [34, 36], Willson was instrumental in the formation of the Acetylene Construction Company (1903) there, which built town-lighting plants in the northwest, and the International Marine Signal Company (1906), which produced a safe, automatic buoy lit with acetylene and employed worldwide [36].

During this same period, Willson had also produced an inexpensive nitrogenous solid from calcium carbide and nitrogen gas that could be ground up for fertilizer, which he hoped to use to revolutionize agriculture [32]. He sold his marine buoy business in 1909 [35] and then sold his Canadian carbide patents to the Canada Carbide Company in 1911 [35, 36]. This later sale was in order to purchase hydroelectric sites on the Shipshaw and Saguenay rivers in Quebec, along with huge timber rights, as he hoped to build a pulp and paper business [36].

Willson required vast amounts of capital to develop his various projects, but had difficulties in finding financial partners. As a result, Willson contracted with Interstate Chemical and James Buchanan Duke (1856–1925), an American tobacco and textile millionaire, to mortgage his fertilizer patents and Quebec properties in 1912, so that he could fund a small fertilizer plant at Lac Meech. Duke had agreed to purchase the operation if he was pleased with the results and within a year the plant was producing beyond expectations. It was at that point, however, that Willson ran short of cash and missed an interest payment, for which Duke seized his assets [34, 36].

Undaunted, Willson found that he still held the rights to carbide production in Newfoundland and Labrador and thus embarked on plans for dams and railways, as well as carbide, pulp and paper, and fertilizer factories that would use the considerable hydroelectric resources there. Had he received the capital promised by British investors or the financial support of the Newfoundland government, Willson might have altered the industrial history of Newfoundland. However, the outbreak of war with Germany prevented the export of British capital [34, 36]. While in New York raising money for yet another project, Willson died of a heart attack on December 20, 1915 [35, 36]. He was later buried in Ottawa [34, 36].

It was on May 2, 1892 in Spray, North Carolina, that Willson accidentally discovered the processes for making calcium carbide and acetylene in commercial quantities [33–36]. This all began with his efforts dating back to 1888 to develop an economical way to make aluminum, which involved the reduction of aluminum ore with carbon in a high-temperature, electric-arc furnace [33, 37, 40], a process explored about the same time in the laboratory of the French chemist Henri Moissan (1852–1907) [35]. In practice, however, Willson was only able to produce a few globules of aluminum in this way. As a solution, he reasoned that if he could first produce a more chemically active metal, such as calcium, he could then use the calcium to reduce the aluminum ore [33, 35]. Thus, he then attempted to produce metallic calcium through the electrothermal reduction of lime (CaO) and various carbon sources [32–34, 37, 38]. For this purpose, he employed an improved electric furnace (Fig. 2.7) with a current of 2000 amperes and 36 V [37, 38], resulting in the production of a dark molten mass that became a heavy, brittle, dark-colored solid upon cooling [38, 39]. Willson is said to have discarded this material, as it was clearly not the metallic calcium sought. It was thus thrown into a neighboring stream, resulting in the liberation of a great quantity of gas [38, 39]. Willson then retained Francis P. Venable (1856–1954), of the University of North Carolina, as a consultant [33, 35]. During the summer and fall of 1892, the process was repeated and refined, during which Venable found the solid substance to be calcium carbide and the gas to be acetylene [32, 33, 35–39]. Willson filed a US patent for the reduction of refractory ores or compounds (including lime)

Fig. 2.7 A schematic of the
original electric-arc furnace
used by Willson Aluminum
Company at Spray, North
Carolina (From Willson and
Suckert [37])

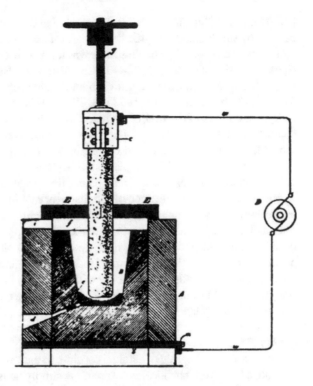

via electric smelting on August 9, 1892 [33, 35, 41], which was granted the following
year on February 21st [41, 42]. Additional US, British, and Canadian patents on the
production of calcium carbide and acetylene followed [34].

Of course, calcium carbide was not a new species and is thought to have been first
prepared in 1839 by Robert Hare (1781–1858) [43], just a few years after the initial
work of Davy on acetylene [8]. Hare had formed calcium cyanide via the reaction of
mercury cyanide with lime, after which he exposed this product to a strong current,
resulting in particles of metallic character that effervesced in water [43, 44]. The first
to recognize the nature of the compound and to identify acetylene as the gaseous
product upon its reaction with water, however, was Friedrich Wöhler (1800–1882)
[42]. It was in 1862 [45] that Wöhler prepared calcium carbide by heating carbon
with an alloy of zinc and calcium, and then reacted it with water to generate calcium
hydroxide and acetylene [37, 38, 45, 46].

The means to make the production of calcium carbide practical, however, did not
come about until after development of electric-arc furnaces capable of providing the
high temperatures needed [39, 46]. William Siemens (1823–1883) first demonstrated
the electric-arc furnace in 1879, with advances allowing higher temperatures by the
1890s, and it was such a modified electric-arc furnace utilized by Willson. Henri
Moissan too had presented a new furnace capable of 3000 °C in December of 1892
and reported its use to generate calcium carbide from quicklime (CaO) and carbon

[47, 48]. L. K. Böhm and L. M. Bullier had also reported the generation of calcium carbide by similar methods [42, 49]. Moissan in particular had communicated to the world his results [47, 48], and thus there was legitimate controversy with regard to the right of priority of discovery [38–40]. Luckily for Willson, he had shared the results of his discovery with Lord Kelvin via a letter and specimen of calcium carbide sent to Glasgow on September 16, 1892 [38, 40]. Kelvin acknowledged the receipt of the sample and his confirmation of its reaction with water in a letter to Willson dated October 3, 1892 [34, 37]. Thus, the issue was ultimately decided in Willson's favor [35, 38, 39], as his documented correspondence with Kelven predated Moissan's December publication. The Imperial High Court of Germany acknowledged this conclusion by annulling the German patent first granted to Bullier [38, 49].

Recognizing the significance of the discovery, Willson continued to develop the technology for the commercial production of calcium carbide using common materials and devoted himself to finding marketable applications [36]. However, after initially failing to find anyone willing to buy their calcium carbide and acetylene patents, Morehead and Willson turned their focus to finding and promoting uses for the manufactured products, beginning with acetylene in lighting. After showing that acetylene could produce a flame 10–12 times brighter than that of coal gas, it was realized that acetylene outshone either coal gas or incandescent electric lights (prior to the development of the tungsten filament) and its use as an illuminant developed rapidly [35, 36].

They made the first sale of calcium carbide, in the amount of one ton, to Eimer and Amend, a New York chemical and apparatus supply house, on January 29, 1894 [33, 35]. Later that same year, they sold their patents for the use of carbide and acetylene in lighting to a new firm, the Electro-Gas Company, of New York City, in August 1894, but retained the rights for chemical manufacturing (Electro-Gas was ultimately absorbed into the Union Carbide Company formed in 1898) [33, 35]. The Electro-Gas Company then made arrangements with the Niagara Falls Power Company to apply 1000 electrical horse-power to the manufacture of calcium carbide, with the plan to increase this to 5000 horse-power shortly thereafter [39]. In turn, the Electro-Gas Company, proceeded to sell carbide manufacturing rights worldwide. However, as part of this agreement, Willson reserved all Canadian rights [33, 35], and Morehead purchased a manufacturing franchise [35].

Moving back to New York in the fall of 1893, Willson set up a laboratory at Eimer and Amend to explore chemical uses for acetylene. After making small quantities of chloroform and various aldehydes, he filed for a patent in February 1894 for the use of acetylene in the manufacture of "hydrocarbon products" [33, 35]. Meanwhile, Morehead completed the first commercial calcium carbide plant in Spray in August 1894 [35]. As publicity concerning the potential of acetylene applications increased, so did the demand for carbide. On May 1, 1895, the plant began to operate around the clock, with the months following providing nothing but continued success [35, 40]. Disaster struck, however, on March 29, 1896, when the plant was destroyed by fire. Morehead, however, built a much larger plant on the James River near Lynchburg, Virginia, which was eventually sold to the Union Carbide Company [35].

With the advent of the robust manufacture of acetylene, significant amounts of the material were now available for a variety of applications [50]. While much of this was focused on acetylene for lighting, its application for oxyacetylene welding and cutting was developed in 1903 [36]. In addition, acetylene became a feedstock chemical for the further production of more complex chemical species [50], as well as a polymerizable monomer for polymeric products.

References

1. Nieuwland JA, Vogt RR (1945) The chemistry of acetylene. Reinhold Publishing Corporation, American Chemical Society Monograph Series, p 1
2. Davy E (1837) Notice of a peculiar compound of carbon and potassium, or carburet of potassium, & c. Br Assoc Adv Sci Rep 5:63–64
3. Davy E (1836) Notice of carburet of potassium, and of a new gaseous bi-carburet of hydrogen. Rec Gen Sci 4:321–323
4. Davy E (1837) Note sur le carbure de potassium, et sur un nouveau bi-carbure d'hydrogène. J Pharm Chim, Ser 2(23):143–322
5. Davy E (1837) Ueber Kohlenstoffkalium und einen neuen Doppelt-Kohlenwasserstoff. Ann Pharm 23:144–146
6. Nieuwland JA, Vogt RR (1945) The chemistry of acetylene. American chemical society monograph series. Reinhold Publishing Corporation, New York, p 5
7. Russell J (1953) Edmund Davy. J Chem Ed 30:302–304
8. Knight D (1992) Humphry Davy. Science and power. Cambridge University Press, Cambridge, pp 127–128
9. Fullmer JZ (2000) Young Humphry Davy. The making of an experimental chemist. American Philosophical Society, Independence Square, PA, p 342
10. Rasmussen SC (2013) It's a Gas! Sir Humphry Davy and his Pneumatic Investigations. In: Patterson G, Rasmussen SC (eds) Characters in chemistry. A celebration of the humanity of chemistry. ACS symposium series 1136. American Chemical Society, Washington, D.C., pp 101–128
11. Davy E (1837) On a new gaseous compound of carbon and hydrogen. Proc R Ir Acad 1:88–89
12. Davy E (1838) On a new gaseous compound of carbon and hydrogen. Br Assoc Adv Sci Rep 6:50
13. Davy E (1839) On a new gaseous compound of carbon and hydrogen. Trans R Ir Acad 18:80–88
14. Rocke A (1985) Chemical atomism in the nineteenth century. From Dalton to Cannizzaro. Ohio State University Press, Columbus, pp 49–97
15. Berthelot M (1860) Sur une nouvelle série de composés organiques, le quadricarbure d'hydrogène et ses dérivés. C R Acad Sci Paris 50:805–808
16. Adloff JP, Kauffman GB (2007) Marcellin Berthelot (1827–1907), chemist, historian, philosopher, and statesman: a retrospective view on the centenary of his death. Chem Educ 12:195–206
17. Doremus CG (1907) Pierre Eugène Marcellin Berthelot. Science 25:592–595
18. Morris PJT (1986) Polymer pioneers. Center for the History of Chemistry, Phiadelphia, pp 30–32
19. Partington JR (1998) A history of chemistry. Martino Publishing, Mansfield Centre, CT 4:465–470
20. Partington JR (1998) A history of chemistry. Martino Publishing, Mansfield Centre, CT 4:355–356
21. Ihde AJ (1964) The development of modern chemistry. Harper and Row, New York, pp 188–189
22. Berthelot M (1863) Recherches sur l'acétylène. Ann Chim 67:52–77

23. Berthelot M (1862) Nouvelles recherches sur la formation des carbures d'hydrogène. C R Acad Sci Paris 54:515–519
24. Berthelot M (1862) Synthèse de l'acétylène par la combinaison directe du carbone avec l'hydrogène. C R Acad Sci Paris 54:640–644
25. Berthelot M (1862) On the synthesis of acetylene by the direct combination of carbon with hydrogen. Chem News 5:184–185
26. Berthelot M (1864) Leçons sur les méthodes générales de synthèse en chimie organique. Gauthier-Villars, Paris, pp 67–84
27. Morren M (1859) De quelques combinaisons gazeuses opérées sous l'influence électrique. C R Acad Sci Paris 48:342
28. Morren M (1862) Synthèse de l'acétylène. C R Acad Sci Paris 55:51–53
29. Berthelot M (1862) Formation de l'acétylène: réponse de M. Berthelot à la Note de M. Morren. C R Acad Sci Paris 55:136
30. Berthelot M (1862) Sur la synthése de l'acétyléne. C R Acad Sci Paris 54:1042–1044
31. Berthelot M (1862) Nouvelles contributions a l'histoire de l'acétyléne. C R Acad Sci Paris 54:1044–1046
32. Berthelot M (1862) Sur la présence et sur le role de l'acétyIene dans la gaz de l'éclairage. C R Acad Sci Paris 54:1070–1072
33. Pratt HT (1976) Thomas Leopold Willson. In: Miles WD (ed) American chemists and chemical engineers. American Chemical Society, Washington DC, pp 512–513
34. Nicholls RVV (1989) Gibbs, LeSueur, and Willson. Pioneers of industrial electrochemistry. In: Stock JT, Orna MV (eds) Electrochemistry, past and present. ACS symposium series 390, American Chemical Society, Washington, DC, pp 525–533
35. American Chemical Society (1998) Discovery of the commercial process for making calcium carbide and acetylene. American Chemical Society, Washington DC
36. Jennifer Paton J (2003) Willson, Thomas Leopold. In dictionary of Canadian biography, vol 14, University of Toronto/Université Laval
37. Willson TL, Suckert JJ (1895) The carbides and acetylene commercially considered. J Franklin Inst 139:321–341
38. Pond GG (1917) Calcium carbide and acetylene, 3rd ed. Bulletin of the Department of Chemistry, Pennsylvania State College, pp 15–16
39. Thompson GF (1898) Acetylene gas, its nature, properties and uses; also calcium carbide, its composition, properties and method of manufacture. Liverpool, pp 21–26
40. Morehead JT, De Chalmot G (1896) The manufacture of Calcium Carbide. J Am Chem Soc 18:311–331
41. Willson TL (1893) Electric reduction of refractory metallic compounds. US Patent 492,377
42. Nieuwland JA, Vogt RR (1945) The chemistry of acetylene. American chemical society monograph series. Reinhold Publishing Corporation, pp 7–10
43. Hare (1839) Process for a fulminating powder—for the evolution of calcium and galvanic ignition of gunpowder. Am J Sci 27:268–270
44. Hare (1840) Calcium. L'Institut 8:310–312
45. Wöhler (1862) Bildung des Acetylens durch Kohlenstoffcalcium. Ann Chem 124:220
46. Ihde AJ (1964) The development of modern chemistry. Harper and Row, New York, p 470
47. Moissan H (1892) Description d'un nouveau four électrique. C R Acad Sci Paris 115:1031–1033
48. Moissan H (1894) Préparation au four électrique d'un carbure de calcium cristallisé; propriétés de ce nouveau corps. C R Acad Sci Paris 118:501–506
49. Böhm LK (1900) Calcium Carbide and Acetylene. Min Ind 9:63–74
50. Morris PJT (1983) The industrial history of acetylene: the rise and fall of a chemical feedstock. Chem Ind 47:710–715

Chapter 3
Cuprene

As access to acetylene became more available, it was only a matter of time until the polymerization of this unsaturated species was investigated. Although the polymerization of acetylene is most commonly associated with the conjugated polymer, polyacetylene, the study of acetylene polymerizations predates the successful formation of polyacetylene by nearly a century. Such early polymerization efforts included both catalyzed and non-catalyzed thermal processes, as well as polymerization via electric discharge, UV irradiation, and α particle irradiation. All of these studies resulted in the production of a resinous material that was eventually given the name *cuprene* [1]. The history of this initial acetylene polymer began in 1866 with the pioneering work of Marcelin Berthelot (1827–1907) [2, 3].

3.1 Berthelot and the Initial Polymerization of Acetylene

Polymerization processes had become a focus for Marcelin Berthelot (Fig. 3.1) by 1863, when he presented his *Lecons sur l'isomérie* before the Paris Chemical Society [4]. In the process, he refined the basic concept of the polymer first introduced by Berzelius in 1832 [5], stating [4]:

> I designate, under the name of polymer bodies, the bodies formed of the same elements, in the same proportion, but under a different state of condensation, and capable of being produced from one another.

© The Author(s) 2018
S. C. Rasmussen, *Acetylene and Its Polymers*, SpringerBriefs in Molecular Science,
https://doi.org/10.1007/978-3-319-95489-9_3

Fig. 3.1 Pierre Eugène
Marcelin Berthelot
(1827–1907) (Courtesy of
Wellcome Library, London,
under Creative Commons
Attribution only licence CC
BY 4.0)

Three years later, he began reporting a series of studies detailing the action of heat upon acetylene [2, 3]. Upon heating acetylene over mercury in a bell at extreme temperatures (described as the softening temperature of glass, or close to the melting temperature)[1] [2, 3], Berthelot observed the formation of a mixture of two products which he described as follows [2]:

> These consist primarily of two carbides: one volatile and which has the properties and reactions of styrene... the other almost fixed, resinous, and which appears to be metastyrol.

The term *metastyrol* was introduced by John Blyth and August Wilhelm von Hofmann (1818–1892) in 1845 to refer to the solid product resulting from the polymerization of styrene (i.e. modern polystyrene) [6]. As such, one could infer that the resinous product described by Berthelot is a polymeric material of some form. The volatile product was described as a yellowish liquid, analysis of which revealed it to consist primarily of benzene, along with styrene and other hydrocarbons [3]. The benzene content was determined to make up ca. 50% of the total products formed in the initial reaction. It is important to note that based on his polymer definition given above, Berthelot considered all of these products (e.g. benzene, styrene, and the resinous solid) to be polymers of acetylene.

---

[1]The glass melting point referred to here would depend completely on the exact type of glass used. As borosilicate glasses were not developed until the late 19th century, the glass here would have been either soda-lime or potash-lime glass, of which soda-lime glass would be the most likely. Even so, the melting point of soda-lime glass could range from 725 to 1000 °C depending on the exact chemical composition [7, 8].

Berthelot continued these studies by heating acetylene in the presence of various solid species, including elemental carbon or iron. It was found that the temperature required to induce reaction was significantly decreased in the presence of these species, while simultaneously increasing the overall reaction rate. In addition, these species also influenced the specific nature of the products generated. He then repeated the process using mixtures of acetylene with equal volumes of either nitrogen, carbon dioxide, methane, or ethane. In each case, he observed the same results although with slower reaction times in comparison to pure acetylene. He ultimately concluded [2]:

> In summary, the transformation of acetylene by heat is not comparable to the phenomena of dissociation: it is not the result of a destruction of the affinity that holds together carbon and hydrogen; but it shall be by a very different mechanism, which is not incompatible with the stability of acetylene. What the heat determines here, it's not a decomposition, it is rather a combination of a higher order, developed by the mutual union of several acetylene molecules.

After these initial reports of Berthelot, additional studies of acetylene polymerization were not reported until the work of Paul Thenard (1819–1884) and Arnould Thenard (1843–1905) in 1874 [9].

3.2 Paul and Arnould Thenard and the Effect of Electric Discharge on Acetylene

Arnould Paul Edmond Thenard[2] was born in Paris in 1819, although sources are less clear on the specific day, with October 6 [10], December 6 [11], and December 16 [12] all cited as the day of his birth. Going by his middle name, Paul was the oldest son, student, and collaborator of the well-known French chemist Louis Jacques Thenard (1777–1857) [10–13]. A man of privilege, Paul was a baron and wealthy landowner in the regions of Côte-d'Or and Saône-et-Loire [10, 11, 13]. However, this also provided him with the time and means for scientific pursuits, particularly in the area of agricultural chemistry. He submitted his first paper to the Paris Academy of Sciences in 1844 [12] and the extent of this work resulted in his election to the Academy in 1864 [10–13], where he became one of its most active members [13]. He was also a member of the French National Society of Agriculture [11].

Within a short span of this 23rd birthday, Paul married Bonne Philippine Isaure Françoise Derrion-Duplan (d. 1921) on October 24, 1842 in Givry, Saône-et-Loire. Three years later, his wife's uncle Pierre Auguste Floret (1785–1847) passed away on January 16, 1847. As he had no children, Floret bequeathed his chateau and land of Talmay, Côte-d'Or, to his niece. The young couple thus moved to Talmay, where Paul installed a laboratory in the vast commons of the chateau [12]. Here, he carried out the bulk of the agricultural chemistry research for which he was known [10–14].

---

[2]Many sources give the family name as Thénard. Partington, however, explicitly states that the name should not be accented [15], nor is it accented in any of the papers of Paul and Arnould Thenard.

Much of this work was motivated by a desire to help the farmers of his vast estates, with particular emphasis on the chemistry of nitrogen and phosphorus [11, 12].

With the outbreak of the Franco-Prussian War[3] in 1870, Paul and his wife initially remained at Talmay to assist with its the defense [11]. On December 2, 1870 [13], however, Paul was taken from his home as a hostage, and transported to Bremen along with several other notables of Côte-d'Or [11–13]. His wife stayed with him during his internment in Bremen [11, 12], where they remained until returning to Talmay at the end of the war in the spring of 1871 [13]. Paul also served the Côte-d'Or commune of Pontailler as general counsel until 1871 and was made a Knight of the National Order of the Legion of Honour [13]. Paul died as a result of apoplexy at his Talmay chateau on August 8, 1884 [10, 14].

Much less is known of the younger Arnould Thenard. He was born in Givry, Saône-et-Loire, to Paul and his wife in 1843 [11], during the first year of their marriage. He was initiated to laboratory life at an early age by his father and he went on to pursue the study of medicine in Paris under the French physician and surgeon Auguste Nélaton[4] (1807–1873) [11]. After his time with Nélaton, he returned to collaborate with his father on various projects. His scientific pursuits were then interrupted by the Franco-Prussian War, during which he joined the French army to serve as a doctor and liaison through enemy lines [11, 12]. After the French defeat in the Battle of Sedan on September 2, 1870, he traveled to Belgium and joined the *Armée de la Loire* [11]. He eventually returned to his scientific work, which spanned chemistry, agriculture, and medicine. Like his father, he was ultimately elected a member of the French National Society of Agriculture [12].

Beginning in 1873, the Thenards began a collaborative effort to study the influence of electric discharge on various mixtures of gases [16, 17]. Such experimental investigations of electric discharge date back to 1857, with a report by Ernst Werner von Siemens (1816–1892) which concentrated on the generation of ozone [18]. In this technique, a flow of gas is subjected to the influence of a discharge across a narrow annular gap between two coaxial glass tubes by an alternating electric field of sufficient amplitude [19]. In this configuration, the electrodes are positioned outside the discharge chamber and thus not in contact with the plasma (Fig. 3.2a). Alternate configurations with only a single dielectric barrier are also known (Fig. 3.2b), in which one electrode would be positioned outside the chamber, while the other would be placed inside the chamber.

---

[3]The Franco-Prussian War, often referred to as the War of 1870 in France, was a conflict between the Second French Empire of Napoleon III and the German states of the North German Confederation led by the Kingdom of Prussia. The war began July 19, 1870 and concluded May 10, 1871.

[4]Auguste Nélaton was born in Paris on June 17, 1807. He studied medicine in Paris, graduating in 1836. He then became *professeur agrégé* in 1839 and professor of clinical surgery in 1851. As a member of the surgical staff of the Hôpital Saint-Louis in Paris, he devised a number of original surgical procedures and developed several phases of plastic surgery. He was elected a member of the Paris Academy of Medicine in 1863 and a member of the French Institute of Science in 1867. The following year he became a Senator of the French Empire. He died in Paris on September 21, 1873 [21].

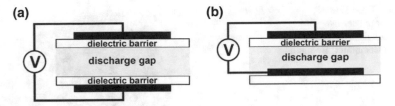

Fig. 3.2 Basic dielectric-barrier discharge configurations

Over time, this electric discharge technique has been referred to by various names, including *silent discharge*, the modern description *dielectric-barrier discharge*, and occasionally *corona discharge* (although this last usually refers to discharges between bare metal electrodes without a dielectric) [19]. Two additional terms, *glow discharge* and *dark discharge* [20], typically refer to whether or not the process generates light, but most often refer to apparatus with direct electrode contact with the gas. The discharge apparatus used in the gas studies of the Thenards are not really described, but what is stated in the discussion of the results appears to be consistent with one of the two configurations given in Fig. 3.2.

Their initial efforts in 1873 were supported by the assistance of Edmond Fremy (1814–1894) and Berthelot in order to study discharge tubes containing mixtures of either marsh gas (i.e. methane) and carbonic acid, or carbon monoxide and hydrogen [16]. This first report was then followed by a second 1873 paper [17] which expanded the number of gases studied to include mixtures of nitrogen and hydrogen or nitrogen and water, as well as the decomposition of species such as water vapor or phosphorous hydride. Following these initial reports, they then published a brief communication in late January of 1874 on the study of acetylene [9].

Using an undescribed discharge device designed by Arnould, they found that the electric discharge caused the acetylene to rapidly condense (4–5 cm^3 min^{-1}), resulting in a solid deposit on the inner walls of the apparatus [9]. They described the deposit as very hard, with a glassy appearance and a color they compared to the dregs of wine (i.e., perhaps burgundy). Analysis of the solid gave an elementary formula identical to that of acetylene gas. Attempts to dissolve the material found it to be insoluble in all solvents investigated and treatment of the solid with fuming nitric acid had no effect. Attempts to dry distill the material also failed, resulting in the conclusion that the solid was analogous or similar to bitumen. This view was shared by Berthelot [9]. By changing the conditions of the experiment, they also produced a liquid body that gave a composition equivalent to acetylene, but could only be produced in very small amounts. As such, the identity was never determined, but these results seem consistent with the previous thermal studies of Berthelot [2, 3].

Although the Thenards never followed up this brief study with further investigation of the product generated, this reported transformation of acetylene has been cited as the earliest known example of the generation of an organic polymer by electric discharge [20]. This report was then quickly followed up with two closely related

Fig. 3.3 P. De Wilde,
professor of chemistry,
Agricultural Institute of the
State in Gembloux, Belgium
[28]

studies. The first of these was by the Belgian P. De Wilde later that same year [22],
with a second study three years later by Berthelot himself in 1877 [23].

3.3 Additional Discharge Studies of Acetylene

Very little is known about the Belgian chemist De Wilde[5] (Fig. 3.3), with even his
given name being unknown beyond the initial "P". What is known is that he was
professor of chemistry at L'Institut Agricole de L'Etat (Agricultural Institute of the
State) in Gembloux, Belgium [24–29]. The Institute had been established in 1860
with G. Michelet as the first professor of chemistry and physics [29]. Michelet was
replaced by De Wilde sometime before 1863 [24], who was then replaced by L.
Chevron in 1867 [28]. While at the Institute, De Wilde taught general chemistry,
analytical chemistry, physics, agricultural technology, and meteorology [28, 29].

De Wilde reported a series of papers on acetylene beginning in 1865, with initial efforts focusing on potential routes to generate acetylene more easily and on
much larger scales. With this goal, he then studied the preparation of acetylene from

---

[5]Additional confusion comes from the fact that De Wilde's name is given inconsistently, with
different sources giving it as "P. De Wilde" [24–28], "P. de Wilde" [29], or "M. P. v. Wilde" [22]. In
this last case, it appears that the honorific title "M." (for Monsieur in French or Meneer in Dutch)
was mistaken as an initial by the German publication. As the first prefix in most common Dutch
names are always capitalized, De Wilde is believed to be the correct name. This is also the form
given in his Belgian papers [24–27] and is the name given by Bulens in his history of L'Institut
Agricole de L'Etat a Gembloux [28].

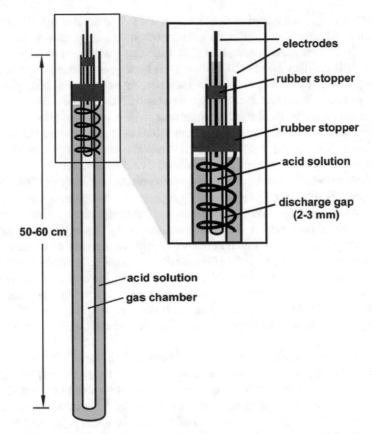

Fig. 3.4 A schematic of De Wilde's electric discharge device based on his written description

ethylene chloride or 1,2-dichloroethane [25, 26]. These efforts did result in the generation of some acetylene, but only in small quantities and as a component of a gaseous mixture. He then continued the following year with research into the reactions of acetylene and hydrogen in the presence of platinum black [27].

Although this early work was limited to the Belgian literature, he then published a paper in *Berichte der Deutschen Chemischen Gesellschaft* in which he first presented a summary of his earlier reports, followed by the results of new studies on the effect of electrical current on various gases and gas mixtures [22]. This paper was published in the spring of 1874 and De Wilde began his description of the electrical experiments with a statement that his experimentation with electric discharge was inspired by the previous work of the Thenards. In comparison to the Thenards, however, De Wilde provided very detailed descriptions of his discharge device (Fig. 3.4), which was most certainly a double dielectric-barrier device consistent with the configuration shown in Fig. 3.2a.

Using the device illustrated in Fig. 3.4, De Wilde first studied the effect of discharge on a 2:1 mixture of sulfur dioxide and oxygen before moving on to pure samples of either ethylene or acetylene [22]. For acetylene, he expected to generate products similar to that found by Berthelot upon heating acetylene at high temperatures [2, 3], citing specifically benzene and styrene, but he states that his experiments did not confirm this expectation. Instead, he observed the condensation of an oily yellow liquid on the walls of the discharge tube, which solidified after a few hours to produce a hard, brittle, material [22]. He went on to characterize the insoluble material as brown in color and amorphous. While the solid could not be dissolved, De Wilde did find that it burned to leave behind a coal-like residue.

De Wilde's statement that he did not generate "polymers of acetylene" ("Polymere des Acetylens") similar to those previously reported by Berthelot is odd and it can only be assumed that he was not aware that Berthelot reported a resinous material in addition to liquid products [2]. It is also interesting that while De Wilde did reference the Thenards' work with ethylene, he did not mention their related work on acetylene that had been published two months previously. De Wilde stated that he intended to continue studying the solid acetylene product, but expected the necessary investigations to be significantly time consuming and thus thought he should first report his initial findings [22]. De Wilde never did follow up on this initial report, however, perhaps because he became aware of the very similar work by the Thenards [9].

Berthelot then followed up the work of both the Thenards [9] and De Wilde [22] with his own study three years later in 1877 [23]. He stated that he was able to verify the accuracy of the Thenards' report, but that he sought to provide some additional detail. Berthelot characterized the solid brown material as a polymer with the formula $(C_4H_2)_n$.[6] Heating a thin layer of the material under N_2 caused it to break down exothermically to give styrene, a carbonaceous residue, and other gaseous byproducts. He stated that this reactivity "distinguishes it from all other known acetylene polymers" [23], but he did not directly compare these results to his previous products from the thermal polymerization of acetylene [2].

After this flurry of acetylene polymerization studies between 1866 and 1877, no further detailed studies appeared for the next 20 years. However, as part of his efforts to commercialize acetylene in the 1890s,[7] Thomas L. Willson (1860–1915) did briefly mention that he observed a "solid poly-acetylene" formed when exposing acetylene to electric spark. Willson stated that the solid resembled horn, and was insoluble in the ordinary solvents [30]. More detailed efforts than began again in 1898 with the report of a new thermal polymerization study by Hugo Erdmann (1862–1910) and Paul Köthner (1870–1932) [31].

---

[6]This is consistent with Berthelot's formula of C_4H_2 for acetylene, see Chap. 2.
[7]See Chap. 2.

3.4 Erdmann and Polymerization Over Copper

Hugo Wilhelm Traugott Erdmann was born on May 8, 1862 in East Prussia [32]. In 1879, at the age of 17, he began the study of chemistry at Halle, Munich and finally Straßburg, where he completed his doctorate in 1883. During his studies, his teachers included Wilhelm Heintz (1817–1880), Adolf von Baeyer (1835–1917), Emil Fischer (1852–1919), and Rudolf Fittig (1835–1910). He then habilitated in 1885 under Jacob Volhard (1834–1910) at Halle, where he became professor in 1894 and the director of the Laboratory of Applied Chemistry in 1899. He was then called to Berlin in 1901 as head of the Laboratory on Inorganic Chemistry of the Institute of Technology. He made significant contributions to both organic and inorganic chemistry, but is perhaps best known for coining the term "noble gases." He died at the relatively young age of 48 during a boating accident on the Müritzsee, a lake in the Mecklenburg region of Germany [32].

In June of 1898, Erdmann and Paul Köthner reported the results of previous studies carried out in 1895 that involved the heating of acetylene over copper metal [31].[8] The report began with the statement that the thermal reaction of pure acetylene occurred at a temperature of 780 °C. It should be noted that this value is in agreement with Berthelot's much less precise report of the temperature of acetylene polymerization [2], but Erdmann and Köthner imply that this primarily resulted in the formation of graphite. They then reported that this temperature could be significantly reduced when carried out in the presence of copper, such that small crystals of graphite could be formed when acetylene was passed over copper powder at 400–500 °C. Furthermore, it was found that if the temperature was maintained below 250 °C, graphite formation was not observed and a light brown solid was produced instead [31]. Continuing the investigation of this brown solid, it was found that it could be produced at a much faster rate, and in larger scale, by using copper oxide in place of copper powder.

Characterization of the light brown material found it to be very light and bulky, with a density of ca. 0.023 g/mL. To test for the presence of copper in the solid, a sample of the material was boiled in dilute HCl and filtered, after which the colorless filtrate was treated with NaOH to precipitate yellow copper hydroxide. The positive detection of copper then led to the conclusion that the solid was a copper compound of some form and combustion analysis suggested a formula of $C_{44}H_{64}Cu_3$. These collected results then led to the conclusion [31]:

> There have been analyzes of different preparations, which give such well-matched values, that we should not hesitate to address this light brown copper acetylene compound as a uniform, albeit very complex composite compound.

Erdmann and Köthner' report was then followed by a closely related conference report in May of the following year by Paul Sabatier (1854–1941) and Jean P. Senderens (1856–1937) [33].

---

[8]In a footnote of the paper, it is stated that a preliminary report of the results was included in Kothner's 1896 dissertation from Halle.

Fig. 3.5 Paul Sabatier
(1854–1941)

3.5 Sabatier and Cuprene

Paul Sabatier (Fig. 3.5) was born on November 5, 1854 at Carcassonne, in the south
of France [34–36]. Initially educated at a lyceum in Carcassonne [35], he moved to
a lyceum in Toulouse in 1868 [36] to prepare for his entrance examinations [35].
He entered the Ecole Normale Supérieure in 1874 [34–36], where he graduated
first in his class in 1877 [35, 36]. After teaching for a year at the Lycée of Nîmes
[35], he became assistant to Marcelin Berthelot at the Collège de France in 1878
[34, 35]. He received the degree of Doctor of Science in 1880, with a thesis on the
thermochemistry of sulfur and the metallic sulfides [34–36].

After a year at Bordeaux in the Faculty of Sciences, he took charge of a course in
physics at the University of Toulouse in January 1882 [34, 35]. He took charge of an
additional course in chemistry in 1883 [35], before becoming professor of chemistry
in November 1884 [35, 36] and then Dean of the Faculty of Sciences in 1905 [34,
36]. Following the deaths of both Henri Moissan (1852–1907) and Berthelot in 1907,
Sabatier was invited to fill their positions at the Sorbonne and the Collége de France,
respectively, but declined both positions to remain in Toulouse [35, 36].

Over his career, Sabatier received many honors. He was awarded the Lacaze prize
of the Paris Académie des Sciences in 1897 [34, 35], was elected a corresponding
member of the Academy in 1901 [35, 36], and awarded its Jecker prize in 1905 [34,
35]. For his method of hydrogenating organic compounds in the presence of finely
disintegrated metals, he was awarded the 1912 Nobel Prize in chemistry, which he
shared with Victor Grignard (1871–1935) [34–36]. The following year, he was made
the first non-resident full member of the Paris Académie des Sciences [35, 36]. In
1915, he received the Davy Medal from the Royal Society [34–36], which elected

him as foreign member in 1918 [34, 35]. In 1933, he received the Franklin Medal from the Franklin Institute in the United States [35, 36].

Sabatier finally retired in 1930 [35, 36]. Although retired, Sabatier received special authorization to continue lecturing, which he did almost to the end of his life [35, 36]. On August 14, 1941, he died in Toulouse at the age of 86 [34–36].

It was at the May 12, 1899, meeting of the Chemical Society of Paris [33], that Sabatier presented initial results obtained with his collaborator Jean Senderens, in which they found that heating acetylene at ca. 180 °C over copper produced a yellow-brown material. This material, which they described as a complex hydrocarbon consisting mainly of ethylenic carbides, was very light and voluminous with small traces of dispersed copper.

This initial report was then followed up with a publication the following year in *Comptes Rendus* [37]. Here, the thermal reaction of acetylene over copper was described in significantly more detail. When a stream of acetylene was passed through a tube containing copper at ambient temperature, no appreciable reaction was observed. However, if the temperature was raised to 180 °C, the copper turned brown and the pressure decreased rapidly due to the observable condensation of acetylene. As this continued, the copper would begin to take on a darker hue and the mass swelled to completely fill the tube, closing off the flow of gas. Interestingly, it was found that if a small amount of the brown substance was smeared into a fresh tube and heated to 180–250 °C under a stream of acetylene, expansion would resume to once again to fill the entire tube. This process could be repeated three to four times before no further reaction commenced [37].

Sabatier and Senderens described the material prepared in this manner as a dark yellow solid, which appears under a microscope to be composed of thin twisted filaments [37]. They went on to say that the material was lightweight and soft, yet a slight compression could give it the consistency and look of wood. It was not found to be soluble in any solvents tested, but it could be burned to give off a smoky flame and aromatic odor, resulting in a black residue. It was finally concluded that the yellow material was a hydrocarbon of empirical formula of C_7H_6, in which small amounts of copper (1.7–3%) were distributed. Due to the origin of the material, they proposed *cuprene* as its name [37].

They admitted that this incomplete study had been published prematurely due to a similar 1899 report by Hans Alexander [38], which they became aware of after Alexander's paper was highlighted in the *Bulletin de la Société Chimique de Paris* in late January 1900 [39]. Still, they maintained their priority of the discovery as their initial Chemical Society of Paris presentation [33] predated Alexander's publication [37]. Of course, it is interesting to note that there is no mention whatsoever of the previous report by Erdmann and Köthner [31]. It may be that they were not aware of it, although as Alexander discusses Erdmann and Köthner's work in his paper [38], it may be that they were conveniently ignoring it to maintain their claim of priority.

3.6   Alexander and More Studies of Acetylene Over Copper

Not much is known about Hans Alexander, other than that he worked in the electrochemical laboratory of the Royal Technical University of Berlin and published a
handful of papers over the time period of 1898–1910. Only one of these papers is
pertinent to the current topic, which he submitted on August 4, 1899 [38], roughly
three months after Sabatier presented his preliminary findings at the meeting of the
Chemical Society of Paris [33]. Alexander begins the paper by highlighting its relationship to the previous work of Erdmann and Köthner [31], as well as some related
work by Sabatier and Senderens on the hydrogenation of acetylene over nickel [40].
Alexander makes no mention of the report of Sabatier at the Chemical Society of
Paris.

Using methods very similar to that of Sabatier and Senderens [37], Alexander
passed acetylene through a tube containing evenly distributed copper. Although no
visible reaction was observed at ambient temperature, slowly heating the acetylene-
filled tube resulted in a change at 225 °C. At this temperature, the gas flow slowed and
the copper began to swell, with greenish, strong-smelling droplets of hydrocarbon
condensing in the colder part of the tube. Further raising the temperature to 260 °C
resulted in the deposition of black shiny carbon crystals on the tube walls. However,
if held at temperatures of 240–250 °C, the reaction progressed smoothly to fill the
whole tube with a light brown mass [38].

The material produced in this way was found to consist of a non-uniform composition. At the entry point of the gas, the material was found to consist of lightweight,
odorless, dark colored flakes. At the other end, however, the material contained a
strong hydrocarbon smell and a slightly darker color. Still, the majority of the material consisted of a uniform, light brown mass. Analysis of the material's copper
content revealed increased copper content in the material near the entry point of the
gas, but the bulk of the material was found to contain ca. 2% copper [38].

As with previous studies, Alexander could not find a solvent capable of dissolving
the material. However, he did find that some copper could be removed by treating
the material with dilute HCl. Still, the copper could not be completely removed by
such treatment, even after boiling the material for several hours. A more successful
treatment was then found in which the material was boiled in HCl containing some
ferric chloride. Treated in this way, the material was now copper free, but did contain
trace amounts of iron (ca. 0.2%), even after boiling with fresh HCl. The nearly metal-
free material appeared a little brighter than the original copper-infused material, but
no other measurable difference was exhibited [38].

Based on these observations, Alexander disputed Erdmann and Köthner's claim
that the material was a copper compound [31] and believed that the copper was only
present in the material as a mechanical mixture. He ultimately concluded [38]:

> In my view, the copper serves only as a contact substance, under the influence of which a
> polymerization of the acetylene takes place. Here, a small amount of aromatic hydrocarbons
> form, which distill out, while mainly a very high molecular weight hydrocarbon of cork-like
> nature arises.

In between the publication of Alexander's 1899 paper and Sabatier's full publication in 1900, yet another related study appeared in November of 1899 [41]. This new study was the first report on the topic from outside Europe, coming from Frank Gooch (1852–1929) and De Forest Baldwin at Yale University in the United States.

3.7 Gooch and Further Studies Over Copper Oxide

Frank Austin Gooch was born in Watertown, Massachusetts [42, 43], on May 2, 1852 [42]. His formal schooling began in Watertown, but at the age of 12, he transferred to Mr. Atkinson's school in Cambridge (later known as the Kendall School) to begin his preparation for college [42]. At the age of 16, he then entered Harvard College in 1868 [42, 43], where he devoted himself to science, especially physics and chemistry [42]. In 1872 [42, 43], he then received the degree of A. B.[9] cum laude, with "*summos in Physicis et Chemia honores*" [42].

Following graduation, Gooch began graduate studies in chemistry, physics, and mineralogy at Harvard [42]. There, he became a teaching assistant to Josiah P. Cooke (1827–1894) in his second year [42, 43]. In 1874, he then served as an assistant in the quantitative analysis laboratory, while also pursuing chemical research under Cooke [42]. His training under Cooke is thought to have been a critical factor in his choice to focus on chemistry for the bulk of his career. At the same time, however, he was also very interested in the physics of crystals, for which he spent the following year studying abroad in Straßburg and Vienna [42, 43]. Gooch returned to Harvard in the autumn of 1876. After completing his A.M. and Ph.D. degrees, he then left again in early June 1877 to pursue the possibility to work with Julius Thomsen (1826–1909) in Copenhagen [42].

His time abroad was short lived, however, and Gooch returned to Harvard to spend two years as private assistant to Wolcott Gibbs (1822–1908) [42, 43]. It was during his time with Gibbs that he developed a new type of filtering crucible that is still known as a Gooch crucible [42]. Following this, he moved to Newport, Rhode Island, where he performed analytical work with the United States Tenth Census (1879–1881) and the North Transcontinental Survey (1881–1884), before moving to Washington to work with the United States Geological Survey (1884–1886) [42, 43]. Then, in 1886, Gooch moved to New Haven, Connecticut, to become professor of chemistry in Yale College, where he remained for the rest of his career [42]. Although his published work covered a range of topics, his efforts focused chiefly in the field of analytical chemistry. He retired in 1918 [43] and died in New Haven on August 12, 1929, at the age of 77 [42].

On November 4th, 1899 [41], Gooch and his coauthor De Forest Baldwin published an extension of the previously discussed study by Erdmann and Köthner [31]. After briefly summarizing the results of Erdmann and Köthner, they state that careful examination of the previous analysis data revealed an error in the calculations [41].

---

[9]*Artium baccalaureus* or a Bachelor of Arts degree.

Gooch and Baldwin thus concluded that the correct carbon to hydrogen ratio should be 6.45:5.70, rather than the inverted relationship represented by the formula of $C_{44}H_{64}Cu_3$ as reported by Erdmann and Köthner. They also expressed doubts about the copper percentage as well, but were unable to provide more accurate values without more detailed information [41].

In order to provide more accurate analytical data and to determine if copper was in fact an integral part of the material generated, the authors then proceeded to perform their own study of acetylene over copper oxide under various conditions [41]. As Erdmann and Köthner had found copper oxide provided faster reaction rates and higher yields compared to copper [31], Gooch was especially interested in the possible role of oxygen in the observed reactions. It was found that the brown solid formed most readily at 225 °C and the application of cuprous or cupric oxide gave nearly identical results. Depending on the specific conditions, the copper content in the product was found to range from 1.54 to 24.21% [41].

To study the effect of copper versus copper oxide, one end of a copper coil was oxidized in a flame. The full coil was then subjected to acetylene under heat, for which product was found to form selectively at the oxidized end, with the other end only changing color [41]. As discussed by Sabatier [37], Gooch and Baldwin also found that the reaction could be initiated by heating fresh acetylene over samples of the brown product alone. Analysis of the resulting products revealed lower percentages of copper in comparison to those generated over the initial copper oxide [41].

As also reported by Alexander [38], Gooch and Baldwin found that in all cases the material produced was not uniform. The bulk of the material was described as a spongy mass of light brown color, but that material closest to the original copper source was darker in color. The material described as of "the brightest color" ("der hellsten Farbe")[10] was found to contain very little copper and no oxygen. In contrast, the darkest product was found to contain higher copper and oxygen content, with the oxygen believed to originate in the initial copper oxide [41]. These observations ultimately led to the same conclusion reached by Alexander [38], that the product was not a copper species and that the copper/copper oxide was simply mechanically entrapped in the hydrocarbon product [41]. Working under the assumption that the product consisted of only carbon and hydrogen, the analytical data was fit to formulas ranging from $C_{12}H_{10}$ to $C_{16}H_{10}$, with an average empirical formula of $C_{14}H_{10}$ [41]. The data of Erdmann and Köthner [31] was then reanalyzed to fit a formula consistent with the low end of this range (i.e. $C_{12}H_{10}$), which is also consistent with the formula determined by Sabatier and Senderens (C_7H_6 or ca. $C_{12}H_{10}$) [37].

Not only did Gooch and Baldwin specifically connect their study to the work of Erdmann and Köthner [31], but they seem to be the first of the researchers discussed so far to directly connect these reactions over copper to the previous thermal polymerizations of Berthelot [2, 3]. As no mention is made of either Sabatier or Alexander, it is unknown if they were aware of these closely related studies. After this second flurry of studies focusing on the reaction of acetylene over copper, no further reports appeared until that of Sima Lozanić (1847–1935) in 1907 [44].

---

[10]A less literal translation could also be "the lightest color" or the "the palest color".

Fig. 3.6   Sima M. Lozanić
(Losanitsch) (1847–1935) in
1905

3.8   Lozanić and a Return to Electric Discharge

Sima M. Lozanić (Losanitsch) (Fig. 3.6) was born in Belgrade, Serbia on February 24, 1847 [45]. He initially studied law at Belgrade College, after which he spent four years studying chemistry under Johannes Wislicenus (1835–1902) in Zürich and August Wilhelm von Hofmann (1818–1892) in Berlin. He returned to Belgrade in 1872 to join the Department of Chemistry at Belgrade College. The College became the University of Belgrade in 1905, with Lozanić appointed the chairman of the University Board. Later that same year, he became the University's first president [45]. He became a corresponding member of the Serbian Academy of Sciences in 1885, only two years after the Academy was founded, and a full member in 1890. He was later elected twice as the president of the Academy in 1899 and 1903. In 1922, Lozanić was awarded an honorary doctorate by the University of Belgrade's Faculty of Philosophy. He retired two years later in 1924, but continued to work until 1929. He died in Belgrade on July 7, 1935, at the age of 88 [45].

In the fall of 1907, Lozanić published a series of experiments on electrosynthesis, in which he subjected various gases or gaseous mixtures to electric discharge [44]. These efforts utilized a discharge device of Berthelot's design, which had been modified to allow him to hermetically enclose the gases in the apparatus, as well as measure the gas pressure during the experiment. In addition, the device could be heated by wrapping it with rubber tubing through which steam was circulated. The gases studied began with either sulfur dioxide, nitric oxide, or their bimolecular

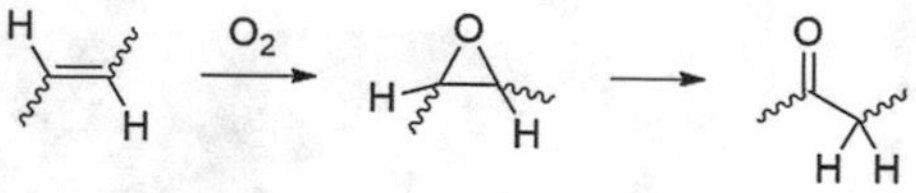

Fig. 3.7 Lozanić's proposed reaction of oxygen with unsaturated units of the acetylene-based solid

mixtures with H_2. The majority of the work, however, focused on acetylene and its bimolecular mixtures with other gases (O_2, CH_4, ethylene, H_2S, CO, and SO_2) [44].

Exposure of pure acetylene gas to electric discharge caused the generation of two products, of which the minor fraction was described as a viscous mass soluble in alcohol or ether. The primary product was an insoluble solid, described as yellow-brown in reflected light and yellow-red in transmitted light [44]. The solid was found to be relatively inert, with no reaction observed upon treatment with hot, fuming nitric acid. Similar to previous observations of Berthelot [23], however, both products rapidly decomposed to carbon residues when heated above 100 °C [44].

Troubled by the fact that analysis of the solid revealed content beyond carbon and hydrogen, Lozanić initially assumed that this was due to an impurity in the acetylene used, but carefully purified acetylene gave similar results [44]. It was then observed that the solid gained mass when stored in a desiccator, with this gain saturated after 26 days at ca. 10%. After eliminating the possibility of nitrogen absorption, it was concluded that this must be due to the absorption of oxygen [44].

Lozanić then followed up the study of this oxygen absorption the following year [46]. The initial solid produced from the exposure of acetylene to the electric discharge was determined to have the formula $C_{48}H_{46}$, which was a little higher in hydrogen in comparison to the equivalent formula (ca. $C_{48}H_{40}$) determined by either Gooch and Baldwin [41] or Sabatier and Senderens [37]. Lozanić then characterized the oxygen uptake of this material and found that it absorbed up to 14 atoms of oxygen without exhausting its saturation capacity. It was concluded that the oxygen absorbed was chemically bound and proposed that the oxygen reacted with carbon-carbon double bonds to form epoxides (Fig. 3.7), which could potentially rearrange to ketone units. As such, he believed [46]:

> From the amount of added oxygen, a conclusion can be drawn on the smallest number of double bonds present.

He ultimately stated that according to their physical and chemical properties, the electrogenerated products apparently belonged to a special class of cyclic, unsaturated compounds. However, he was less certain about the overall structure of these materials, warning [46]:

> On the constitution of the electrocondensed hydrocarbons still very little can be said at present with certainty.

Although it appears that Lozanić was familiar with previous reports of Berthelot, he did not mention the previous electric discharge studies of either the Thernards or

Fig. 3.8 Daniel Paul Alfred
Berthelot (1865–1927)
(Courtesy of Wellcome
Library, London, under
Creative Commons
Attribution only licence CC
BY 4.0)

De Wilde, nor did he connect the products of electric discharge to those of thermal polymerization. After Lozanić's study of the oxygen absorption, no further reports appeared until that of Daniel Berthelot (1865–1927) two years later in 1910 [47].

3.9 Daniel Berthelot and UV Polymerization

Daniel Paul Alfred Berthelot (Fig. 3.8) was born in Paris on November 8, 1865 [48, 50], the son of Marcelin Berthelot [48, 49]. The younger Berthelot was educated at the Sorbonne and the Muséum national d'Histoire naturelle, his teachers including Paul-Quentin Desains (1817–1885), Henri Becquerel (1852–1908), and Gabriel Lippmann (1845–1921) [49]. He became assistant preparator at the laboratory of physics research of the Faculty of Sciences of Paris in 1884. In 1892, he was appointed assistant to Becquerel, who had just become the chair of physics at the Muséum national d'Histoire naturelle [50]. Two years later, he became an associate at the l'Ecole supérieure de Pharmacie in 1894 and was then appointed professor of physics in 1903 [48–50].

At the l'Ecole supérieure de Pharmacie and in his laboratory of plant physics at Meudon [48, 49], he became known for his work in physical chemistry, including efforts in pyrometry, the electrolytic nature of acids, the physical characterization of gases and photochemistry. For his accomplishments, the Académie des Sciences awarded Berthelot the Jecker Prize in 1898 and Hughes Prize in 1906 [48]. He was elected a member of the Academy of Medicine in 1914 [50] and a member of the

Académie des Sciences in 1919 [48, 50]. Berthelot died suddenly on March 8, 1927, at the age of 62 [48, 50].

In 1910, Berthelot and his coauthor Henri Gaudechon published the first report of the photochemical polymerization of acetylene [47]. Using a quartz mercury vapor lamp (110 volts, 2.5 amps), acetylene gas in a quartz tube was irradiated with UV light for one hour to generate a solid product on the tube walls. Due to the non-transparency of the product, it was reported that the reaction was quite rapid at the start, but slowed over time. The solid generated was described as yellow in color, with the characteristic odor of acetylene polymers generated via electric discharge. No benzene was produced in the process and any residual gas was determined to be unreacted acetylene [47].

In order to investigate the polymerization at lower pressures of acetylene, irradiation of acetylene mixed with either hydrogen or nitrogen was then studied. This was stated to provide more efficient generation of the yellow solid, with the secondary gases appearing to act only as inert buffer gases. Finally, irradiation of acetylene-ethylene mixtures was investigated, resulting in a yellow solid identical in appearance to the previous acetylene polymers, along with a greasy coating concluded to be condensed ethylene [47]. Unfortunately, no further characterization of the yellow solid products was reported.

Although the comment concerning acetylene polymers generated via electric discharge makes it appear that the authors were familiar with such previous studies, no particular studies were mentioned or referenced, nor was any connection made to any of the previous thermal polymerization studies, including those of his father. Following the report of Berthelot and Gaudechon, there was another sizeable gap in related studies. However, efforts ramped up again in the 1920s and 30s, beginning with the work of H. P. Kaufmann (1889–1971) in 1918 [51].

3.10 Kaufmann and Discharge Versus Thermal Polymerization

Hans Paul Kaufmann was born in Frankfurt, Germany, on October 20, 1889, [52]. Beginning in 1908, he studied chemistry in Heidelberg, Berlin, and Jena. He became an assistant at the University of Jena's Chemical Institute in 1911 and obtained his Ph.D. there under Ludwig Knorr (1859–1921) in January 1912 [52, 53]. He then continued at the Chemical Institute until 1914, when he joined the German army with the outbreak of World War I [52, 53]. After the delay caused by the war, Kaufmann obtained his habilitation at Jena on May 17, 1916 [51–53], while on leave from military duty [52]. Seriously wounded shortly thereafter, he was assigned to war-related scientific work following his recuperation [52].

In 1919, Kaufmann became a. o. professor[11] [52, 53] and Director of the Analytical Division of the Chemistry Institute at Jena [52]. After the early death of his mentor Knorr in 1921, he moved to the Pharmaceutical Institute at Jena [52]. There, he completed the pharmaceutical state examination [53] and began teaching in 1922 [52]. In 1931, he became professor of pharmacy at the University of Münster [52, 53], before moving to Berlin as professor of pharmaceutical chemistry in 1943 [52]. He returned to Münster as professor of pharmacy and chemical technology in 1946, becoming professor emeritus in 1958, but continued as the director of the Pharmacy and Food Chemistry Institute until April 1959 [52]. Kaufmann died October 2, 1971, after an extended illness [52].

Although best known for his work on fats and oils [52, 53], his habilitation thesis focused on the polymerization of acetylene [51, 52]. Published in 1918 [51], his habilitation research attempted to provide some clarity on the composition of the products generated by subjecting acetylene to electric discharge. Kaufmann began by reviewing the previous discharge work of the Thenards [17], De Wilde [22], Berthelot [23], and Losanitsch (Lozanić) [44, 46]. He then began his efforts by optimizing the reaction conditions in order to obtain a suitable amount of the products for study. In the process, he found that under normal conditions, the discharge device generated heat and such conditions favored the formation of solid products. However, if the apparatus was suitably cooled, only the liquid product was produced [51].

Kaufmann also found that the addition of ice water to the discharge tube caused the product firmly adhered to the apparatus walls to loosen as a brown, brittle mass [51]. Even better, he found that if the oil produced under cooling was then subjected to further discharge as a heated solution, a fine, light yellow powder was produced. This powder exhibited all of the properties of the normally generated solid mass, but was more suitable for further analysis. Kaufmann then proceeded to study this powder via treatment with various chemical reagents [51].

The investigation of such chemical reactions began with the study of oxygen absorption, which Lozanić had viewed as the addition of oxygen across double bonds and thus the amount of oxygen absorbed should allow calculation of the number of double bonds contained in the electrogenerated product [46]. Kaufmann disagreed, however, stating that the nature of oxygen binding could not be inferred from the material available at the time [51]. Continuing his study, Kaufmann then treated the solid product with various reducing and oxidizing agents. No reactions were observed with reducing agents, but nitrated products were generated by heating the solid material in dilute nitric acid for prolonged periods of time. Greater reactivity was observed upon treating the solid with alkaline potassium permanganate, ultimately giving low amounts of benzoic, isophthalic, and terephthalic acids. As a result, Kaufmann concluded that much of the material structure consisted of unsaturated chains, which generated carbon dioxide upon oxidation by the permanganate [51].

Kaufmann then followed this report with two additional papers in the early 1920s [54, 55]. These additional papers contrasted acetylene polymerization via electric

---

[11]*Außerordentlicher professor* or *professor extraordinarius* is a professor without a chair and typically works under a chaired professor.

discharge to its catalyzed and non-catalyzed thermal polymerization [54] and probed the copper-catalyzed process in more detail [55]. As previously noted by Erdmann and Köthner [31], as well as Gooch and Baldwin [41], Kaufmann recognized that the nature of the catalyst played a significant role, with copper bronze and copper oxides more effective than pure copper [54]. He also showed that cuprene could be successfully produced using catalytic cupriferrocyanide ($CuFe(CN)_6^{2-}$) [54], but ultimately concluded that the presence of oxygen was a critical factor [55].

As shown previously by Alexander [38], Kaufmann found that the copper content could not be removed by treatment with HCl. Boiling the solid in aqua regia for several hours, however, removed all but a trace amount of copper [54]. As with his previous study of the electric discharge products [51], analysis of the copper-catalyzed product by reduction was unsuccessful. Treatment of the products with 80% HNO_3, however, gave mellitic acid (benzenehexacarboxylic acid), benzoic acid, and a naphthalene derivative as oxidation products [54]. Lastly, it was found that treatment with Br_2 in the presence of iron halides gave brominated products, although the exact nature of the products depended on the specific conditions used [54].

These collective results led Kaufmann to conclude that cuprene was not a uniform substance, and that the material must comprise a mixture of acetylenic condensation products, the composition of which was variable [55]. In addition, he stated that cuprene must contain benzene units, as well as attached carbon chains that were neither completely saturated, nor fully unsaturated [54], and felt that the material's unsaturation was primarily present as aromatic units [55]. Two years later, the Belgian Walter Mund (1892–1956) showed that the application of alpha rays could also cause acetylene polymerization [56].

3.11 Mund and Polymerization via Alpha Particles

Walter Emile Marie Mund was born on January 22, 1892, in Antwerp, Belgium [57]. After his primary and secondary education at the Jesuit College in Antwerp, he enrolled as a candidate in the natural sciences at the University of Leuven (Louvain in French; Löwen in German) in October of 1910. There, he worked in the laboratory of physical chemistry under the direction of Pierre Bruylants (1885–1950). He was awarded his doctorate in natural sciences on July 24, 1914, after defending a dissertation entitled "On the Vapor Pressures of Sulfur Dioxide" [57].

Within days of completed his doctorate, World War I broke out, interrupting what should have been the start of a promising academic career. Thus, Leuven fell to the German First Army in August of 1914, and was the subject of mass destruction shortly thereafter. Mund and his family were evacuated to Manchester, England, where he worked as a chemist for the Southern Cotton Oil Company [57]. It is unclear, however, whether Mund left Leuven before or after the German occupation. In early 1916, Mund left his position to join the Belgian Army at the Yser Front, first as a mere rifleman, and later a corporal. His actions during the war earned him the Cross of Fire and the Medal of Victory [57].

Following the war, Bruylants became chair of general chemistry at Leuven and thus Mund was called to become his successor to lead the laboratory of physical chemistry [57]. For more than 30 years he taught physical chemistry and distinguished himself by research in what would now be described as radiation chemistry, with particular emphasis on the study of chemical reactions induced by α particles. He was a member of the Royal Academy of Belgium, the Chemical Society of Belgium, and the Faraday Society. Mund died August 15, 1956, at the age of 64 [57].

In early 1925, Mund and W. Koch reported a study on the effect of radiation on various hydrocarbon gases, including methane, ethane, ethylene, and acetylene [56]. While the effect on the saturated gases was minimal, greater effect was observed for ethylene and acetylene. In the latter case, a 40 cm^3 bulb of acetylene was treated with 58 millicuries of emanation. The reaction was monitored for 8 days, with a dense fog filling the bulb within the first day, resulting in the deposition of a fine yellow-brown powder. They described the collected yellow powder as odorless, light as pollen, and with no obvious crystalline structure when examined under a microscope. The product exhibited low solubility in both xylol and ether, with no trace of reaction in sulfuric acid up to 300 °C. It was ultimately concluded that the product was a hydrocarbon with a formula approximating that of acetylene [56] and proposed that it could be identical to the material previously reported by Alexander [38] via copper-mediated thermal conditions.

Mund and Koch published a second paper later that same year, in which the effect of α particles on acetylene was studied in more detail [58]. Here, the number of α particles emitted by the radon source was determined against a radium reference. Under these conditions, they were able to determine that each α particle caused the condensation of 4.38×10^6 acetylene molecules. They admitted, however, that not all of the acetylene molecules may have undergone polymerization and some could have been absorbed by the resulting solid produced. In terms of the material's structure or the mechanism of its formation, they admitted that neither could be easily determined. However, they were suitably convinced that the material was identical to cuprene [58]. These investigations were continued the following year with a focus on the effect of oxygen, pressure, and temperature [59]. It was concluded, however, that none of these variables had any effect on the previously determined values.

Mund later returned to the study of acetylene polymerization via α particles with two more papers in the mid-1930s [60, 61]. The first of these discussed acetylene polymerizations in the presence of CO_2 [60], while the second focused on the formation of benzene during polymerization [61]. However, by this time the primary study of acetylene polymerization had been continued by Samuel Lind (1879–1965) [62].

3.12   Lind and the Polymerization of Acetylene Under
Various Conditions

Samuel Colville Lind was born June 15, 1879, in McMinnville, Tennessee [63]. Initially educated in the McMinnville public schools through age 16, Lind enrolled at
Washington and Lee University of Lexington, Virginia in 1895. The school empha-
sized the classics and he spent his first three years studying primarily language and
culture (French, Latin, Greek, German, and Anglo-Saxon). Entering his final year,
he still needed six credits in the sciences, and he was persuaded to take chemistry to
meet this requirement. Although he had very little previous chemistry knowledge,
he became captivated by the subject due to the influence of James (Jas.) Lewis Howe
(1859–1955) [64], professor of chemistry and head of the Chemistry Department.
Lind received his B.A. in 1899 and then returned to Washington and Lee for additional courses in chemistry, geology and mineralogy [63].

   Lind then entered the Massachusetts Institute of Technology (MIT) in the fall
of 1902 for additional coursework [63]. MIT did not give graduate degrees at the
time, but he was able to pursue research under Arthur Amos Noyes (1866–1936).
In 1903, he was then awarded a Dalton traveling fellowship that allowed him to go
to the Institut für Physikalische Chemie in Leipzig. There, he began research under
Max Bodenstein (1871–1942) on the reaction kinetics of H_2 and Br_2. Receiving his
Ph.D. in August 1905, Lind was then offered an assistantship by Bodenstein. but he
decided to return to the United States to accept a teaching position at the University
of Michigan in September 1905 [63].

   At Michigan, Lind was in charge of the physical chemistry teaching laboratory
and he was unable to carry out much research. This changed in 1910, when he spent
time in the Paris laboratories of Marie Curie (1867–1934) gaining proficiency in the
handling of radioactive species [63]. The following year, he then moved to the newly
formed Institut für Radiumforschung in Vienna, where he studied the action of α
particles on oxygen molecules. Such study of chemical reactions caused by ionizing
radiation then became his main field of research for the remainder of his career [63].

   In 1913, Lind accepted an appointment with the U.S. Bureau of Mines, where
he worked on radium extraction from carnotite. In 1925, he then became assistant
director of the Fixed Nitrogen Research Laboratory of the U.S. Department of Agriculture. He did not stay long, however, becoming the head of the School of Chemistry
at the University of Minnesota in 1926. He remained at Minnesota until his retirement in 1947, after which he became a consultant to the Union Carbide Corporation
in 1948. He later also served as acting director of Oak Ridge National Laboratory's
chemistry division for several years. He passed away on February 12, 1965, while
fishing below Norris Dam in Tennessee [63].

   In 1924, while still at the U.S. Department of Agriculture, Lind began studying
the effect of ionizing radiation on ethane, before expanding these efforts to additional
gases (methane, ethane, propane, butane, ethylene, acetylene, $(CN)_2$, HCN, and NH_3)
the following year [62]. In terms of acetylene, Lind confirmed the previous results of
Mund and Koch [58] on pure acetylene, while also reporting the polymerization of

acetylene-N_2 mixtures. In both cases, the same solid yellow powder was produced, which he directly compared to cuprene as previously reported by Alexander [38] and Sabatier [37].

Lind focused his efforts to just organic unsaturated gases (acetylene, $(CN)_2$, HCN, ethylene) the following year [65]. In this second paper, he again confirmed that the exposure of acetylene to α radiation resulted in a light-yellow powder, which he stated was similar to materials generated via copper-catalyzed reactions by Sabatier and Senderens [37], Alexander [38], and Kaufmann [54, 55], as well as related materials produced via UV light by D. Berthelot [47] or even electric discharge [65]. For the most part, his acetylene results were very similar to those previously reported by Mund and Koch [58], although he does propose that cuprene results from the polymerization of 20 acetylene molecules, thus giving the structure $(C_2H_2)_{20}$ [65].

Following his move to Minnesota in 1926, Lind then shifted to studying the parameters of UV-induced polymerization of acetylene in an attempt to compare the polymerization via α particles to that of UV light [66, 67]. He began the initial 1930 communication with the statement that the results suggested that the polymerization rate was proportional to the intensity of absorbed light but is otherwise independent of the acetylene pressure. In the process, he also found that wavelengths shorter than 253.7 nm were required to cause polymerization, most likely due to acetylene's transparency at longer wavelengths. Lastly, he determined the quantum yield for the photoinduced reaction to be 7.4 ± 2.5 [66].

Lind published the corresponding full paper in 1932 [67], which provided the full experimental details, but essentially came to the same conclusions given in the previous communication. Lind did report, however, that UV irradiation of cuprene films under vacuum revealed no measurable change, thus eliminating the possibility of significant secondary reactions of cuprene. In addition, he was able to provide a more accurate value of 9.2 ± 1.5 for the quantum yield of the polymerization [67].

Lind published a final note on the photochemical polymerization in 1934 [68], primarily to refine the quantum yield to account for the generation of ethylene and ethane as byproducts. As a result, the total quantum yield for all photochemical processes was estimated to be 9.7. Based upon previous results, he then proposed the composition of UV-generated cuprene to be in the range of $(C_{10}H_9)_n$ to $(C_{15}H_{14})_n$, which is in good agreement with the formulas given by Gooch and Baldwin [41] or Sabatier and Senderens [37] for the copper-catalyzed thermal products and close to his own previous idealized formula of $(C_2H_2)_{20}$ [65]. To explain the reduced hydrogen content, he then proposed the following radical mechanism for the reaction:

$$C_2H_2 + h\nu \rightarrow C_2H_2{}^* \rightarrow {}^\bullet C_2H + {}^\bullet H$$
$${}^\bullet C_2H + C_2H_2 \rightarrow {}^\bullet C_4H_3$$
$${}^\bullet C_4H_3 + C_2H_2 \rightarrow {}^\bullet C_6H_5 \text{ (etc.)}$$
$${}^\bullet C_nH_{n-1} + {}^\bullet C_mH_{m-1} \rightarrow \text{solid}$$

While his focus in the 1930s was on the UV polymerization of acetylene, Lind also attempted to compare the reactions caused by the exposure of gases to electric discharge to those caused by α particles [69, 70]. This began with a study of methane,

ethane, propane, butane, and ethylene in 1930 [69], followed by a second paper in 1931 that focused on methane, ethylene, and acetylene [70]. The emphasis of this second report was on the reaction rate and potential insight into the reaction mechanism, but no firm conclusions could be made and it was determined that further analytical data were required.

Lind's final acetylene paper was a 1937 study of the oxygen absorption of cuprene prepared by α particle irradiation [71]. After production of the solid polymer, the reaction vessel was evacuated and filled with a known pressure of oxygen at room temperature, after which the reaction was monitored via changes in the gas pressure. In contrast to the proposed reactions of Lozanić [46], Lind concluded that the majority of the oxygen was adding to the cuprene, but that a small amount of CO was also lost. Analysis of the oxidized product revealed that it contained ca. 25% oxygen, leading to the following proposed reaction

$$(C_2H_2)_{20} + 5^1/_2O_2 \rightarrow C_{39}H_{40}O_{10} + CO$$

3.13   What Exactly Is Cuprene?

By the end of the 1930s, interest in cuprene was in decline [1]. Of those additional papers still being published, these could be split primarily into two groups; either continued study on acetylene polymerization via radiation or efforts to determine the structure of cuprene and the mechanism of its polymerization. The conclusion of this chapter will endeavor to summarize the results of this last group of publications.

By the early 1920s, it had become clear that cuprene was not a copper-based species and contained no copper when fully purified. Still, as noted by Kaufmann, the name cuprene was retained even though it had been shown to be a hydrocarbon [54]. At least one contributing factor was that cuprene was the only name ever proposed for the material, with all other references being the "acetylene polymer" or even more general references. Within this same time frame, it was also starting to become clear that all of the various polymerization methods described above were producing the same or nearly the same material, with an empirical formula very close to that of acetylene. Still, the structure of this polymer was unknown.

The mechanism proposed by Lind in 1934 [65] is essentially an addition polymerization,[12] the product of which should be the linear polymer now recognized as polyacetylene, $(HC=CH)_n$. In fact, this was actually thought to be the structure of cuprene for a while, with even such polymer luminaries as Paul Flory (1910–1985) reporting its structure as this [72]. However, this structure is not consistent with the results of Kaufmann [54, 55], who viewed cuprene as consisting of cyclic structures and at least partially saturated units. In addition, although Lind believed cuprene to exist in repeat lengths of ca. 20 [65], the brownish-yellow color of cuprene (Fig. 3.9)

---

[12]See Chap. 1.

Fig. 3.9 Sample of a
cuprene film generated via
ion irradiation (Courtesy of
Reggie Hudson)

was not consistent with polyenes longer than 4–9 repeat units.[13] Of course, the substantiated report of true polyacetylene by Natta in the late 1950s [73–75] then finally confirmed that cuprene did not contain a linear conjugated structure.

As early as 1937 [76], it was proposed that the initial acetylene polymerization was indeed a chain growth addition process. However, as the linear chain being produced was still unsaturated, various secondary polymerization processes could occur to result in the observed cuprene product. This view was further supported by the fact that cuprene was completely insoluble and no solvent was ever found that even induced swelling of the material [76].

Further support for this proposed polymerization mechanism was then given in 1964, when a paper reported that linear polyacetylene could be converted to a cuprene-like material when heated in air at temperatures above 200 °C [77]. As a result, it was thus proposed that the formation of cuprene proceeds via the initial polymerization of acetylene into linear polyenes, which were then converted to the final yellow-brown product by the action of residual oxygen. This proposed mechanism would also be consistent with the earlier conclusions of Gooch that oxygen played an important role in the formation of cuprene [41].

This mechanism was then further reinforced in 1971 by a detailed study of cuprene generated via γ-ray irradiation [78]. Examination of the cuprene product with an electron microscope revealed that it consisted of fairly uniform spheres with a diameter of ca. 3000 Å. From the size, it was estimated that each particle contained 2.3×10^8 molecules of acetylene. Combining the observations of their study with the results of previous studies, the authors then proposed the following multi-stage mechanism of cuprene formation [78]:

1. Primary acetylene polymerization to form polyenes of ca. C_4H_4 to $C_{20}H_{20}$.
2. Secondary polymerization of the initially generated polyenes.
3. Physical condensation of the polyenes of ca. $C_{20}H_{20}$ generated in steps 1 and 2 to form liquid spherical particles.
4. Rapid and extensive crosslinking/polymerization of the polyenes within the droplets to form the final solid cuprene particles.

---

[13] See Chap. 4.

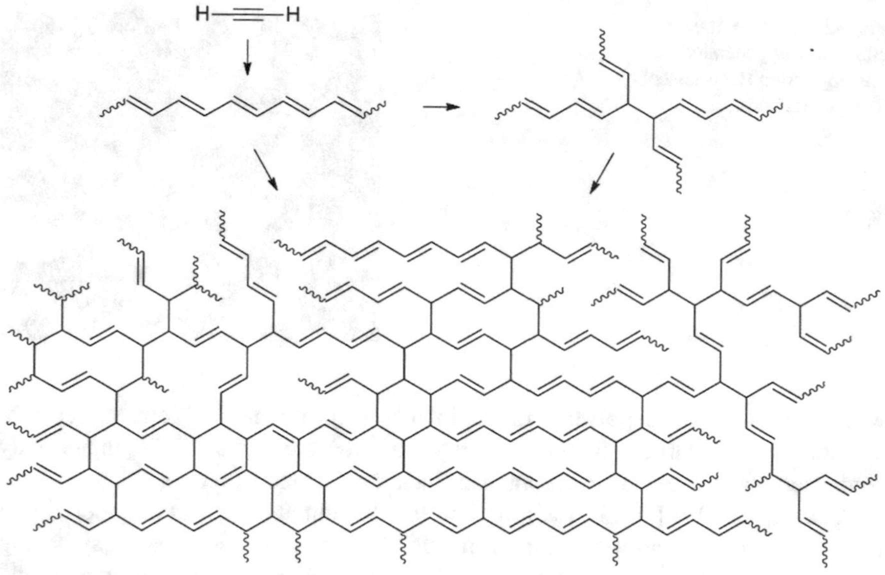

Fig. 3.10 A simplified representation of the proposed formation of cuprene

A simplified representation of this process is shown in Fig. 3.10. However, it must be stressed that the intractability of cuprene, coupled with its likely structural complexity, has made detailed determination of this structure impossible. As such, there is still much unknown about both its absolute structure and the full mechanistic details of its formation.

The product generated via the mechanism given above would then consist of both saturated and unsaturated sections, of which the unsaturated sections would consist of relatively low conjugation lengths, thus giving cuprene its yellow color. As such, cuprene could be viewed as a heavily crosslinked polyacetylene. However, as the crosslinking removes points of unsaturation, this may more closely resemble a crosslinked polyethylene with some limited points of unsaturation [1]. This latter view is of historical interest as crosslinked polyethylene was not reported until 1953 [79].

References

1. Rasmussen SC (2017) Cuprene: a historical curiosity along the path to polyacetylene. Bull Hist Chem 42:63–78
2. Berthelot M (1866) Action de la chaleur sur quelques carbures d'hydrogène. (Première partie). C R Acad Sci Paris 62:905–910
3. Berthelot M (1866) Les polymères de l'acétylène. Première partie: synthèse de la benzine. C R Acad Sci Paris 63:479–484
4. Berthelot M (1866) Lecons sur l'isomérie. L. Hachette et Compagnie, Paris, p 18

5. Berzelius JJ (1833) Isomerie, Unterscheidung von damit analogen Verhältnissen. Jahresber Fortschr Phys Wiss 12:63–67
6. Blyth J, Hoffman AW (1845) Ueber das Styrol und einige seiner Zersetzungsproducte. Ann Chem Pharm 53:289–329
7. Rasmussen SC (2012) How glass changed the world. The history and chemistry of glass from antiquity to the 13th century. Springer Briefs in Molecular Science: History of Chemistry, Springer, Heidelberg, pp 2–4
8. Rasmussen SC (2015) Modern materials in antiquity: an early history of the art and technology of glass. In: Rasmussen SC (ed) Chemical technology in antiquity. ACS symposium series 1211, American Chemical Society, Washington, D.C., Chapter 10, pp 267–313
9. Thenard P, Thenard A (1874) Acétylène liquéfié et solidifié sous l'influence de l'effluve électrique. C R Acad Sci Paris 78:219
10. Feddersen BW, von Oettingen AJ (eds) (1898) J. C. Poggendorff's Biographisch-Literarisches Handwörterbuch zur Geschichte der Exacten Wissenschaften, vol 3. J. A. Barth, Leipzig, p 1334
11. Liébaut M (1907) Une famille d'illustres savants: quelques souvenirs de Conté, Humblot-Conté, Jacques, Paul et Arnold Thenard. Bull Soc Encour Ind Natl 106:675–698
12. Bouchard G (1950) Avant-propos. In: Thenard P (ed) Un Grand Français: Le Chimiste Thenard, 1777–1857. Imprimerie Jobard, Dijon, pp 5–8
13. Vapereau G (1880) Dictionnaire Universel des Contemporains: Contenant Toutes les Personnes Notables de la France et des Pays Étrangers, 5th edn. Hachette, Paris, pp 1731–1732
14. Anonymous (1884) Science Gossip. The Athenaeum 2966 (Aug. 30):282
15. Partington JR (1998) A history of chemistry, vol 4. Martino Publishing, Mansfield Centre, CT, p 90
16. Thenard P, Thenard A (1873) Sur les combinaisons formées sous l'influence de l'effluve électrique par le gaz des marais et l'acide carbonique d'une part, et l'oxyde de carbone et l'hydrogène d'autre part. C R Acad Sci Paris 76:1048–1051
17. Thenard P, Thenard A (1873) Nouvelles recherches sur l'effluve électrique. C R Acad Sci Paris 76:1508–1514
18. Siemens W (1857) Ueber die elektrostatische Induction und die Verzögerung des Stroms in Flaschendrähten. Ann Phys Chem 102:66–122
19. Kogelschatz U (2003) Dielectric-barrier discharges: their history, discharge physics, and industrial applications. Plasma Chem Plasma Process 23:1–46
20. Kolotyrkin VM, Gil'man AB, Tsapuk AK (1967) Production of organic ourface oilms by the action of electrons, ultraviolet radiation, and the glow discharge. Russ Chem Rev 36:579–591
21. Walsh JJ (1911) Auguste Nélaton. In The Catholic Encyclopedia. Robert Appleton Company, New York
22. v. Wilde MP (1874) Vermischte Mittheilungen. Ber Dtsch Chem Ges 7:352–357
23. Berthelot M (1877) Quatrième mémoire. Absorption de l'hydrogène libre par l'influence de l'effluve. Ann Chim Phys 10:66–69
24. De Wilde P (1863) De l'action du protochlorure de phosphore sure l'acide monochloracetique.—Nouveau mode de preparation du chlorure d'acétyle chloré. Bull Acad R Belg 16(ser 2):487–489
25. De Wilde P (1865) Sur la production de l'acétylène.—Nouvelles méthodes. Bull Acad R Belg 19(ser 2):12–15
26. De Wilde P (1865) Sur la production de l'acétylène.—Nouvelles méthodes. Bull Acad R Belg 19(ser 2):91–94
27. De Wilde P (1866) Action de l'hydrogène sur l'acétylène sous l'influence du noir de platine. Bull Acad R Belg 21(ser 2):31–32
28. Bulens C (ed) (1910) L'Institut Agricole de L'Etat a Gembloux, 1860–1910. Imprimerie Scientifique, Brussels, pp 79, 101–102, 111–112
29. Bulens C (ed) (1904) The Government Agricultural Institute of Gembloux. Imprimerie Scientifique, Brussels, pp 7–18, 27-30

30. Willson TL, Suckert JJ (1895) The carbides and acetylene commercially considered. J Franklin Inst 139:321–341
31. Erdmann H, Köthner P (1898) Einige Beobachtungen über Acetylen und dessen Derivate. Z Anorg Chem 18:48–58
32. Lockemann G (1959) Erdmann, Hugo Wilhelm Traugott. In Neue Deutsche Biographie, Duncker & Humblot, Berlin 4:572
33. Hanriot M (1899) Extrait des Procès-verbaux des séances. Bull Soc Chim Fr 21:529–531
34. Rideal EK (1942) Paul Sabatier, 1859–1941. Biogr Mem Fellows R Soc 4:63–66
35. Taylor HS (1944) Paul Sabatier, 1859–1941. J Am Chem Soc 66:1615–1617
36. Morachevskii AG (2004) Paul Sabatier (to 150th Anniversary of His Birthday). Russ J Appl Chem 77:1909–1912
37. Sabatier P, Senderens JB (1900) Action du cuivre sur l'acétylène: formation d'un hydrocarbure très condensé, le cuprène. C R Acad Sci Paris 130:250–252
38. Alexander H (1899) Ueber die Einwirkung des Acetylens auf Kupfer. Ber Dtsch Chem Ges 32:2381–2384
39. Auger V (1900) Action de l'acétylène sur le cuivre: Hans Alexander. Bull Soc Chim 24:38
40. Sabatier P, Senderens JB (1899) Hydrogénation de l'acétylène en présence du nickel. C R Acad Sci Paris 128:1173–1176
41. Gooch FA, Baldwin DF (1899) Die Einwirkung von Acetylen auf die Oxyde des Kupfers. Z Anorg Chem 22:235–240
42. Van Name RG (1931) Frank Austin Gooch. Biogr Mem Nat Acad Sci USA 15:105–135
43. Browning PE (1923) Frank Austin Gooch. Ind Eng Chem 15:1088–1089
44. Losanitsch SM (1907) Über die Elektrosynthesen. II. Ber Dtsch Chem Ges 40:4656–4666
45. Snežana B (2006) Sima Lozanić (1847-1935). In: Djordjević VD, Vitorović D, Marinković D (eds) Lives and work of Serbian scientists. Serbian Academy of Sciences and Arts, Belgrade, pp 153–155
46. Losanitsch SM (1908) Die Sauerstoffabsorption der elektrokondensierten Körper. Monatsh Chem 29:753–762
47. Berthelot D, Gaudechon H (1910) Effets chimiques des rayons ultraviolets sur les corps gazeux. Actions de polymérisation. C R Acad Sci Paris 150:1169–1172
48. Wisniak J (2010) Daniel Berthelot. Part I. Contribution to thermodynamics. Educ quím 21:155–162
49. Barrois C (1927) Mémoires et Communications des Membres et des Correspondants de l'Académie. C R Acad Sci Paris 184:637–641
50. Boutaric A (1927) Daniel Berthelot. Rev Scientifique 65:353–357
51. Kaufmann HP (1918) Über die Produkte der Einwirkung der dunklen elektrischen Entladung auf Acetylen. Ann Chem 417:34–59
52. Knothe G (2004) Giants of the past. Hans Paul Kaufmann (1889–1971). Inform 15:802–803
53. Täufel K (1959) Professor Dr. Dr. h. c. H. P. Kaufmann. Nahrung 3:789–792
54. Kaufmann HP, Schneider M (1922) Acetylen-Kondensationen, I.: Versuche zur Konstitutionsermittlung des Cuprens. Ber Dtsch Chem Ges 55:267–282
55. Kaufmann HP, Mohnhaupt W (1923) Zur Theorie der Bildung des Cuprens (Acetylen-Kondensationen, II.). Ber Dtsch Chem Ges 56:2533–2536
56. Mund W, Koch W (1925) Sur l'altération chimique de quelques hydrocarbures gazeux sous l'action du rayonnement radioactif. Bull Soc Chim Belg 34:119–126
57. Belche R (1956) A la mémoire de Walter Mund. Inst Grand-Ducal Luxembourg Sect sci nat phys et math Arch 23:23–25
58. Mund W, Koch W (1925) Sur la polymérisation de l'acétylène sous l'action des particules α. Bull Soc Chim Belg 34:241–255
59. Mund W, Koch W (1926) The chemical action of α particles on acetylene. J Phys Chem 30:289–293
60. Mund W (1934) The insert gas effect in the radiochemical polymerization of acetylene. J Phys Chem 38:635–637

61. Mund W (1937) The formation of benzene in the radiochemical polymerization of acetylene. J Phys Chem 41:469–475
62. Lind SC, Bardwell DC (1925) The chemical effects in ionized organic gases. Science 62:422–424
63. Laidler KJ (1998) Samuel Colville Lind. Biogr Mem Natl Acad Sci USA 74:227–242
64. Kauffman GB (1972) The work of James Lewis Howe. Bibliographer Platin Group Met 16:140–144
65. Lind SC, Bardwell DC, Perry JH (1926) The chemical action of gaseous ions produced by alpha particles VII. Unsaturated carbon compounds. J Am Chem Soc 48:1556–1575
66. Lind SC, Livingston RS (1930) The photochemical polymerization of acetylene. J Am Chem Soc 52:4613–4614
67. Lind SC, Livingston RS (1932) The photochemical polymerization of acetylene. J Am Chem Soc 54:94–106
68. Lind SC, Livingston RS (1934) The photochemical polymerization of acetylene. J Am Chem Soc 56:1550–1551
69. Lind SC, Glockler G (1930) V. The condensation of hydrocarbons by electrical discharge. Comparison with condensation by alpha rays. J Am Chem Soc 52:4450–4461
70. Lind SC, Schultze GR (1931) The condensation of hydrocarbons by electrical discharge. VIII. The condensation as a function of time and pressure. J Am Chem Soc 53:3355–3366
71. Lind SC, Schifle CH (1937) Studies in the oxidation of alpha ray cuprene. J Am Chem Soc 59:411–413
72. Flory PJ (1937) The heat of combustion and structure of cuprene. J Am Chem Soc 59:1149–1150
73. Natta G, Pino P, Mazzanti G (1955) Polimeri ad elevato peso molecolore degli idrocarburi acetilenicie procedimento per la loro preparozione. Italian Patent 530,753 (July 15, 1955); Chem Abst 1958, 52:15128b
74. Natta G, Mazzanti G, Pino P (1957) Hochpolymere von Acetylen-Kohlenwasserstoffen, erhalten mittels Organometall-Komplexen von Zwischenschalenelementen als Katalysatoren. Angew Chem 69:685–686
75. Natta G, Mazzanti G, Corradini P (1958) Polimerizzazione stereospecifica dell'acetilene. Atti Accad Naz Lincei Rend Cl Sci Fis Mat Nat 25:3–12
76. Calhoun JM (1937) A study of cuprene formation. Can J Research 15b:208–223
77. Fredericks RJ, Lynch DG, Daniels WE (1964) Thermal properties of polyacetylene: on the origin of cuprene in Reppe's cyclooctatetraene synthesis. J Polym Sci Part B Polym Lett 2:803–808
78. Briggs JP, Back RA (1971) The effect of electric fields on the radiolysis of acetylene, and the mechanism of cuprene formation. Can J Chem 49:3789–3794
79. Morawetz H (1985) Polymers: the origins and growth of a science. Wiley, New York, pp 180–181

Chapter 4
Polyenes and Polyvinylenes

Although not produced via the direct polymerization of acetylene, the following chapter will introduce the history of two earlier materials structurally related to the later formal polyacetylene [1, 2]. These two classes of materials were known as *polyenes* and *polyvinylenes*, two terms that at times have been used synonymously with polyacetylene. Like polyacetylene, both polyenes and polyvinylenes can be represented by the idealized formula $(-CH=CH-)_n$, yet with important differences [1–6]. Polyenes were the simpler of the two, but were typically limited to shorter oligomers ($n = 2$–10) [1, 2]. In contrast, polyvinylenes were polymeric analogues generated via the condensation of HX species from substituted polyethylenes [e.g. poly(vinyl halide)s or poly(vinyl alcohol)]. As a result, the resulting polyvinylenes were typically limited by the introduction various defects from incomplete elimination [1].

4.1 Polyenes

Polyenes are formally defined as materials containing an even number of methyne ($=CH-$) groups covalently bonded to form a linear carbon chain bearing one p electron on each carbon atom [3–6]. As such, the chemical structure of polyenes is best represented by a formula $R-(CH=CH)_n-R$, where n denotes the number of repeating units and R can represent a variety of capping groups. For the current presentation, the discussion will be limited to simple examples of symmetrically end-capped polyenes. The study of polyenes dates back to at least 1928, particularly with the concentrated efforts of Richard Kuhn (1900–1967) [7–10].

© The Author(s) 2018
S. C. Rasmussen, *Acetylene and Its Polymers*, SpringerBriefs in Molecular Science,
https://doi.org/10.1007/978-3-319-95489-9_4

4.1.1   Richard Kuhn and the Study of Conjugated Double Bonds

Richard Kuhn was born December 3, 1900, in Döbling, a suburb of Vienna, Austria [11, 12]. He was educated at the Döblinger Gymnasium and was introduced to chemistry at an early age by Ernst Ludwig, a family friend and director of the Institute of Medicinal Chemistry [11, 12]. At age 17, he was conscripted into the army during World War I and served in the Telegraph Regiment beginning in March 1918 [12]. With the end of conflict, he was discharged on November 18, 1918, and he enrolled at the University of Vienna only days later [11, 12]. He only stayed for two semesters, however, before moving to Munich. There he completed his undergraduate studies and was accepted for doctoral work under Richard Willstätter (1872–1942) in 1921. He was awarded his Ph.D. in November of 1922 with a dissertation on the specificity of enzymes in carbohydrate metabolism. He continued research under Willstätter, completing his habilitation in March 3, 1925 [11, 12].

Kuhn lectured on chemistry at Munich until 1926, when he was called to Eidgenössische Technische Hochschule in Zurich as full professor of general chemistry and head of the analytical laboratory [11, 12]. In May of 1928 [12], he then accepted an invitation to become director of the Institute of Chemistry at the newly founded Kaiser Wilhelm Institut für Medizinisch Forschung[1] at Heidelberg, a position which began April 1, 1929 [12]. At the same time, he was appointed honorary professor in the Faculty of Mathematics and Natural Sciences of the University of Heidelberg [11, 12]. Following the death of the Institute's founder Ludolf von Krehl in 1937, Kuhn became Director of the Institute, a position he held until the end of his life. In 1950, he also became professor of biochemistry in the Faculty of Medicine at the University of Heidelberg [11, 12].

Kuhn had a distinguished career and his application of chemical analysis to biological topics during the 1930s contributed strongly to the early development of modern biochemistry. For work on the synthesis and analysis of polyenes and vitamins, he was awarded the 1938 Nobel Prize in Chemistry, with many additional awards and honors to follow. He served as President of the Deutsche Chemische Gesellschaft from 1938-1945 and of the Gesellschaft Deutscher Chemiker from 1964 to 1965 [11]. Kuhn's health began to fail in 1965 and he died of cancer on July 31, 1967 [11, 12].

Beginning in 1928, while still in Zurich, Kuhn and Alfred Winterstein published a series of papers on the topic of conjugated double bonds, focusing on the synthesis and study of polyenes [7–10]. This began with the synthesis of a series of phenyl-capped polyenes, $Ph–(CH=CH)_n–Ph$ (where $n = 1$–8) [7, 13]. Kuhn reasoned that the phenyl groups were necessary to prohibit the polyenes from undergoing secondary polymerization reactions. While the first two members of this series, stilbene (**I**) and 1,4-diphenylbutadiene (**II**), where well known at the time, the triene (**III**) and tetraene (**IV**) analogues had only been recently reported and the higher members had been unknown before Kuhn and Winterstein's 1928 report [7].

---

[1] It was renamed the Max Plank Institut in 1950.

These synthetic efforts began with optimizing the synthesis of 1,6-diphenyl hexa-triene (III) in order to generate the polyene in greater quantities [7]. As illustrated in Fig. 4.1, the reductive dimerization of cinnamic aldehyde generated the correspond-ing diol intermediate. This intermediate was then treated with phosphorus diiodide (P_2I_4) to convert the diol to the diiodo derivative, which is unstable at room temper-ature and eliminates I_2 to give the desired hexatriene (III).

Kuhn and Winterstein then proceeded to optimize the synthesis of 1,8-diphenyl octatetraene (IV) [7]. As shown in Fig. 4.1, this involved the double condensation of cinnamic aldehyde with succinic acid in the presence of lead(II) oxide, with acetic anhydride as the solvent. Here, the lead oxide was found to be critical to obtain reasonable yields. Other metal oxides were also investigated, but lead oxide was the only species found to have the proper solubility under the reaction conditions.

With the optimization of 1,8-diphenyloctatetraene (IV) in hand, this synthetic methodology was used as a general framework to access higher polyenes via the utilization of larger conjugated aldehydes (Fig. 4.2) [7, 13]. In this way, polyenes with an even number of double bonds could be obtained as evidenced by the isolation of 1,12-diphenyldodecahexaene (VI) and 1,16-diphenylhexadecanoctaene (VIII). To access the polyenes with an odd number of double bonds, succinic acid was replaced with dihydromuconic acid as shown in Fig. 4.2 [7, 13]. While this allowed the isolation of 1,10-diphenyldecapentaene (V) and 1,14-diphenyltetradecaseptaene (VII), the production of 1,18-diphenyloctadecanonaene was unsuccessful [7].

Kuhn and Winterstein then went on to characterize the diphenylpolyene series, with selected results summarized in Table 4.1. In particular, they were interested in the fact that while multiple isomers were known for stilbene and diphenylbutadiene, the higher polyenes were all isolated as a single isomeric species [7]. This fact, coupled with the regular trend in the corresponding melting point, led to the conclusion that all of the higher polyenes were isolated as the all-*trans* form. Trends in the colors of the various polyenes were also investigated under various conditions [7, 10] and the series was later characterized by X-ray crystallography in 1930 [14].

Kuhn, Karl Hausser, and coauthors then characterized the absorption [15, 16] and fluorescence [17] properties of the series in 1935. In the case of the absorption data, measurements were performed both at room temperature [15] and low temperature

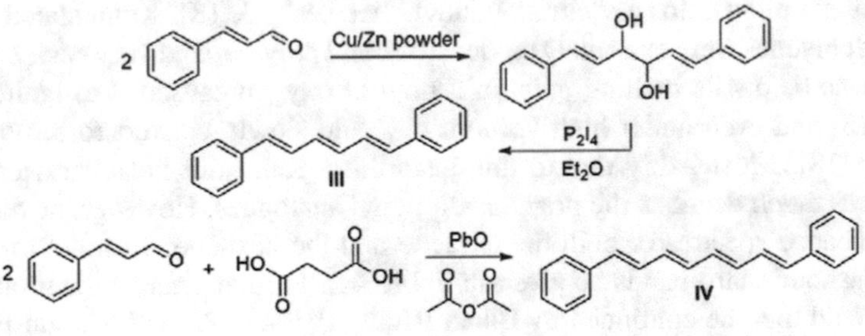

Fig. 4.1 Kuhn's synthesis of 1,6-diphenylhexatriene (III) and 1,8-diphenyloctatetraene (IV)

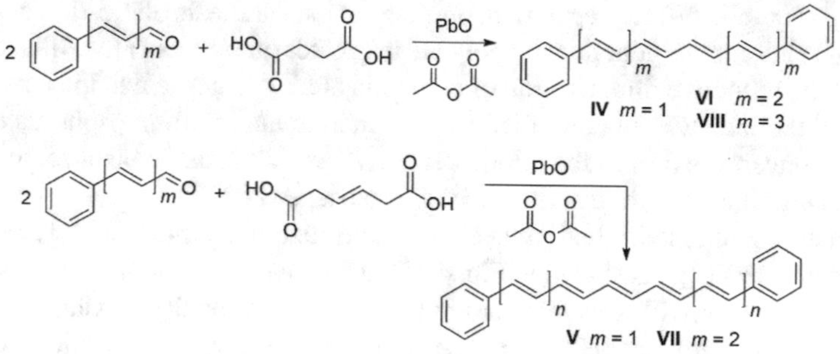

Fig. 4.2 Kuhn's general synthesis of higher diphenylpolyenes **IV–VIII**

Table 4.1 Selected properties of diphenylpolyenes, Ph–(CH=CH)$_n$–Ph ($n = 1$–8) [7]

Polyene	n	Total possible isomers	Melting point (°C)	Polyene	n	Total possible isomers	Melting point (°C)
I	1	2	124[a]	**V**	5	20	253
II	2	3	152.5[b]	**VI**	6	36	267
III	3	6	200	**VII**	7	72	279
IV	4	10	232	**VIII**	8	136	285

[a]*Trans* form of **I**; [b]*Trans-trans* form of **II**

[16]. For the room temperature data (Table 4.2), it was found that for the lowest energy absorption, both the value of the λ_{max} and the absorption intensities exhibited a near linear dependence on the number of double bonds (n) in the polyene [15].

Kuhn went on to synthesize and study polyene series end-capped with other functional groups beyond phenyl [13]. For the current discussion, the most relevant of these where the methyl-capped series CH$_3$–(CH=CH)$_n$–CH$_3$ ($n = 1$–4, 6) [13, 18], which was the first series of polyenes without conjugated end-groups. As shown in Fig. 4.3, these were prepared by treating the known polyenealdehydes with ethylmagnesium bromide to generate the polyeneol intermediate, which could be easily dehydrated via treatment with a 2% solution of p-toluenesulfonic acid in ether [18].

The first four members of this series were colorless and it was only the hexaene that exhibited color as "citrate-yellow" needles [13, 18]. Kuhn stated that the dimethylpolyenes were extremely air-sensitive and polymerized very easily. Heating the tetraene below its melting point in a flow of oxygen caused it to ignite explosively [13] and even under high vacuum it would slowly convert to a rubber-like polymer [18]. Most likely due to this instability, Kuhn did not characterize this series in as much detail as the previous diphenyl analogues. However, he did report the absorbance spectra for both the tetraene and the hexaene, with both members exhibiting four sharp bands as given in Table 4.3. Further characterization of this series would then be continued by Elkan Blout (1919–2006) and Melvin Fields in 1948 [19, 20].

Table 4.2 Absorption properties of diphenylpolyenes, Ph–(CH=CH)$_n$–Ph ($n = 1$–7)[a] [15]

Polyene	n	Band 1		Band 2		Band 3	
		λ_{max} (nm)	ε_{max} ($\times 10^{-3}$)	λ_{max} (nm)	ε_{max} ($\times 10^{-3}$)	λ_{max} (nm)	ε_{max} ($\times 10^{-3}$)
I	1	319[b]	50[b]	306[b]	56	294[b]	54[b]
II	2	352[b]	60[b]	334	92	316[b]	70[b]
III	3	377	120	358	172	343	125
IV	4	404	177	384	198	363	134
V	5	424	204	403	216	387	140
VI	6	445	250	420	261	400	176
VII	7	465	280	435	310	413	200

[a]In benzene; [b]Values are uncertain as the bands are not very pronounced

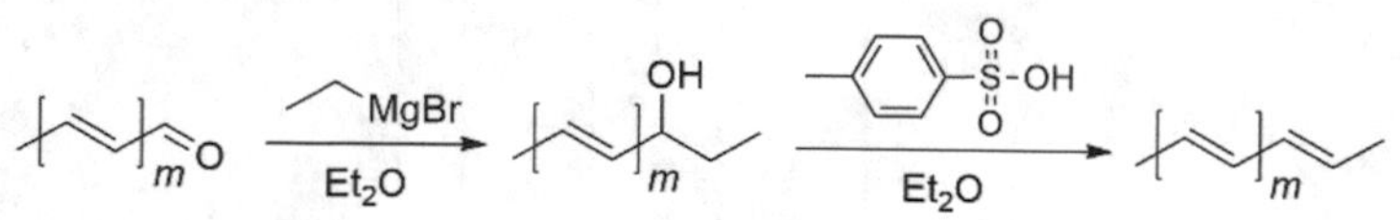

Fig. 4.3 Kuhn's general synthesis of dimethylpolyenes

Table 4.3 Absorption data for dimethylpolyenes, $CH_3-(CH=CH)_n-CH_3$ ($n = 4,6$) [18]

n	λ_{max} (abs, nm)	λ_{max} (abs, nm)[a]	λ_{max} (abs, nm)	λ_{max} (abs, nm)
4[b]	320	297	283	272
6[c]	375	360	340	328

[a]Most intense peak in the spectrum; [b]In hexane; [c]In chloroform

4.1.2 Blout and the Spectroscopy of Polyenes

Elkan Rogers Blout was born in New York City on July 2, 1919 [21]. Initially educated at DeWitt Clinton High School, he then enrolled at Philips Exeter Academy as he was still too young to attend college when he graduated [21]. After a year at Exeter, Blout then attended Princeton University. Completing his B.A. in chemistry in 1939, he attended Columbia University, where he received his Ph.D. in 1942 [21, 22]. He then worked as a research fellow with Louis Feiser and R. B. Woodward at Harvard University before being offered a position at the Polaroid Company in 1943 [21]. At Polaroid, he helped develop the instant photographic process and the color translating microscope. Blout left Polaroid in 1962 to become professor of biological chemistry at Harvard Medical School [21, 22]. He was awarded the National Medal of Science in 1990 [21, 22] and the American Chemical Society's Ralph F. Hirschmann Award in Peptide Chemistry in 1991 [21]. Blout died from pneumonia on December 20, 2006 [22].

In January of 1948, Blout and his coauthors began to report studies on the UV-visible and infrared (IR) spectra of various polyenes [19, 20]. While the majority of the polyenes investigated were asymmetric polyenes terminated with aldehydes and either methyl or furyl groups, the characterization of the symmetrical dimethyl polyenes was also reported (Table 4.4) [19]. As with the previous diphenyl series, a near linear relationship was found between the energy of absorption and the number of double bonds (n), although the linear relationship could be improved by plotting the square of the absorption energy verses n [19]. In terms of the IR spectra, characteristic strong C–H frequencies were found at 1445–1455 cm^{-1}, as well as several other bands of lower intensities [20]. The strongest of the bands that could possibly be ascribed to double bond vibrations was found near 1650 cm^{-1}.

In the process of these studies, Blout and his coworkers also recognized that these polyenes could have possible charged resonance forms (Fig. 4.4), in which charges could exist at different lengths along the conjugated backbone [19]. This could be viewed as a prelude to the eventual understanding of mobile charges in

Table 4.4 Spectroscopic data for various dimethylpolyenes, $CH_3-(CH=CH)_n-CH_3$ [19, 20]

n	UV-vis absorbance (λ_{max}, nm)[a]	Infrared frequencies (cm^{-1})[b]
2	227	1445, 1608, 1656, 1693, 1805
3	263	1445, 1601, 1646, 1685, 1804
4	295	1450, 1615, 1646, 1677,1825
6	358[b]	1444, 1567, 1611, 1635, 1664[c]

[a]In hexane. The maxima reported are for the most intense band, rather than the lowest energy band;
[b]In chloroform; [c]Solid-state film

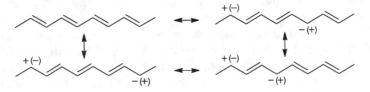

Fig. 4.4 Resonance forms of 2,4,6,8-decatetraene ($n = 4$)

conjugated systems. Further efforts on both diphenyl- and dimethyl-polyenes were then continued by Ferdinand Bohlmann (1921–1991) in 1952 [23].

4.1.3 Bohlmann and Higher Dimethylpolyenes

Ferdinand Bohlmann was born August 28, 1921 in Oldenburg, Germany. He began his education in Oldenburg, before entering Göttingen in 1839 at the age of 18 [24]. There he studied chemistry and despite some interruptions by war service, he completed his chemistry diploma in 1944. He then continued his studies under Karl Dimroth (1910–1995). Bohlmann is said to have followed after Dimroth moved to the University of Marburg in 1944 [24], but he still received his doctorate from Göttingen in 1946 with a thesis on the solvatochromism in the pyridine series [25].[2] Bohlmann then continued at Marburg, working with Hans Herloff Inhoffen (1906–1992), who he followed to Braunschweig in 1947. He completed his habilitation at Braunschweig in 1952. Bohlmann then became a lecturer and in 1957, a non-chaired professor. Finally, in 1959, he became the successor of Friedrich Weygand (1911–1969) at the Institute of Organic Chemistry at the Technical University of Berlin, where he focused on the isolation, structure elucidation and synthesis of natural products. He died on September 23, 1991 [24].

In January of 1952, while still at Braunschweig, Bohlmann reported a new synthetic route to symmetrically end-capped polyenes, with its demonstrated use for

---

[2]Some sources state that his doctoral research was under Hans Brockmann (1903–1988), Director of the Institute for Organic Chemistry at Göttingen. However, Bohlmann's thesis topic is very much in line with the research focus of Dimroth and the work appears to have been under his direction.

74   4 Polyenes and Polyvinylenes

the production of both diphenyl- and dimethyl-polyenes [23]. As shown in Fig. 4.5, this started with the application of the previous methods of Karrer and Coehand [26] for the production of symmetrical polyene-diones. Bohlmann then presented a new efficient and effective method for the conversion of the dione intermediate to diol via LiAlH4 reduction. Finally, the diol could be transformed into the final polyene through the use of P2I4 as previously developed by Kuhn and Winterstein [7].

Of the four polyenes Bohlmann reported to demonstrate his new methods, both of the diphenylpolyenes where already known species. More critically, however, he generated two new members of the dimethylpolyene series; 1,12-dimethyldodecahexaene consisting of five double bonds and 1,18-dimethyloctadecanonaene with nine double bonds, the highest polyene reported for the time [23]. This record polyene was isolated as orange-yellow crystals that were extremely oxygen sensitive and oxidized rapidly. While this limited analysis, Bohlmann did report its UV-visible spectrum in benzene, which exhibited exceptional fine structure consisting of seven bands, with the most intense band at ca. 410 nm and the lowest energy band at ca. 450 nm.

Bohlmann then returned to the synthesis of higher dimethylpolyenes in 1956 to further extend the series up to ten double bonds [27]. This work was based upon the goal of developing synthetic methods that would allow the generation of a pure all-*trans* species in the final step and thus not require chromatography of the product. Towards this goal, he decided that a suitable solution was the partial hydrogenation of acetylene or diacetylene-containing polyenes. The *cis*-alkene segments produced by the hydrogenation would provide higher solubility, such that isomerization to the poorly soluble all-*trans* products could then allow separation from the more soluble isomers via crystallization. To accomplish this, he constructed polyenes containing central alkyne or dialkyne units via the recently reported methods of Georg Wittig[3] [28], as shown in Fig. 4.6. The purified alkyne intermediate was then reduced to the corresponding *cis*-alkene using Lindlar catalyst and hydrogen, after which the polyene was isomerized with a trace of iodine to give the desired all-*trans* product.

As expected, the higher polyenes were found to be easily oxidized, although this was only a significant problem at elevated temperatures. As such, no defined melting

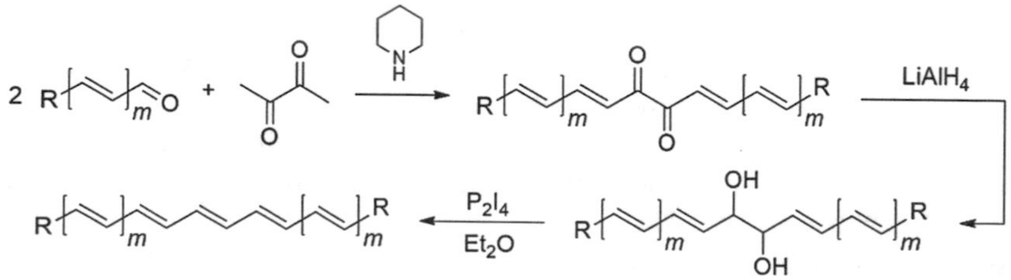

Fig. 4.5 Bohlmann's new route to polyenes

---

[3]This is now known as the Wittig reaction or Wittig olefination, for which Wittig was awarded the Nobel Prize in Chemistry in 1979.

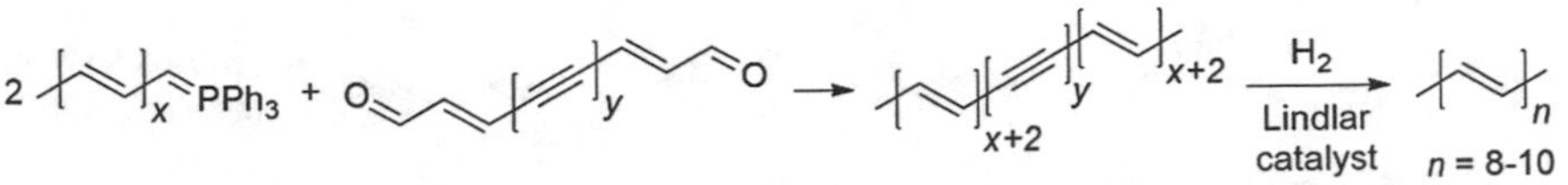

Fig. 4.6 Bohlmann's synthesis of higher dimethylpolyenes

Table 4.5 UV-visible data for the dimethylpolyene series, $CH_3-(CH=CH)_n-CH_3$ [27]

n	λ_{max} (nm)[a]	n	λ_{max} (nm)[a]
2	227	6	352
3	263	8	395.5
4	299	9	412.5
5	326	10	419 (432[b])

[a]In ether. The maxima reported are for the most intense band, rather than the lowest energy band;
[b]Value for the dihexyl analogue, $H_{13}C_6-(CH=CH)_{10}-C_6H_{13}$

points could be determined, even in the absence of oxygen, with all three polyenes decomposing when heated. Although the higher all-*trans* dimethylpolyenes exhibited limited solubility, dilute solutions could be produced to allow measurement of their absorption spectra (Table 4.5). The highest member of this series, however, provided such low solubility that it was difficult to acquire trustworthy spectroscopic data. To overcome this limitation, Bohlmann synthesized a functionalized decaene analogue containing hexyl endgroups, which exhibited enhanced solubility and provided better absorption data (Table 4.5) [27].

4.1.4 Unfunctionalized Polyenes

As might be expected, the development of the synthesis and characterization of the fully unfunctionalized parent polyenes occurred later than the diphenyl- and dimethyl-analogues discussed above. These efforts really began with the report of high purity 1,3,5-hexatriene by G. Forrest Woods and Louis H. Schwartzman in 1948 [29], along with 1,3,5,7-octatetraene a year later [30]. As outlined in Fig. 4.7, the initial step utilized the same methods as applied by Kuhn for the synthesis of dimethylpolyenes, after which the polyen-ol intermediates were dehydrated under N_2 with activated alumina at elevated temperature. While the triene was a liquid at room temperature, the tetraene was a white, crystalline solid that could be recrystallized from petroleum ether. The pure tetraene, however, was reported to be quite air sensitive, giving off marked white fumes and a pungent aldehyde odor [30]. Because of its rapid decomposition, an accurate melting point could not be determined. The UV absorption of both compounds was also reported (Table 4.6) [29, 30].

The next two members of the series were then reported in 1952 by Alexander D. Mebane [31]. As shown in Fig. 4.8, this involved reduction of either a carboxylic

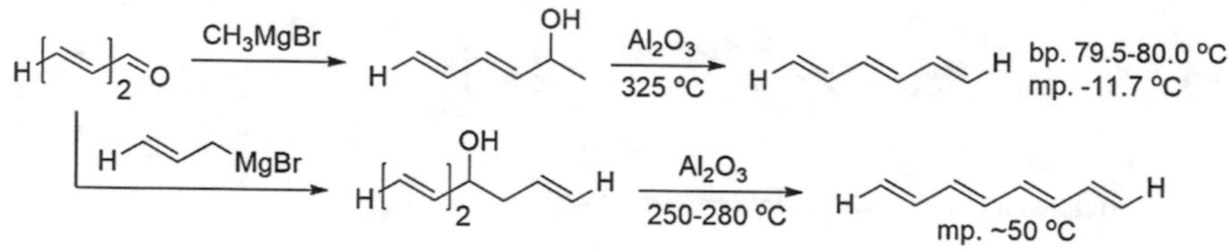

Fig. 4.7 Woods and Schwartzman's synthesis of unfunctionalized polyenes

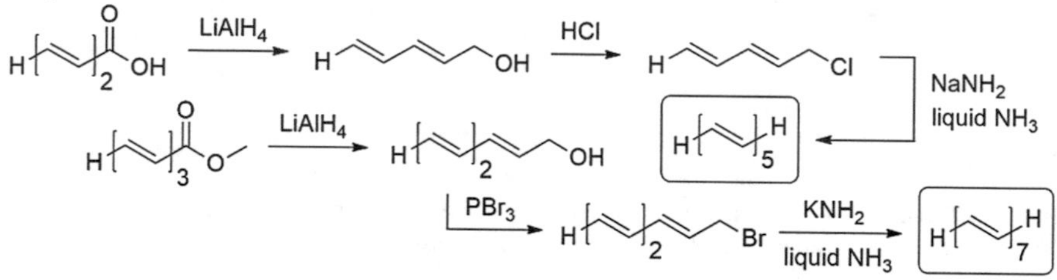

Fig. 4.8 Mebane's synthesis of unfunctionalized polyenes

acid or ester terminated polyene, followed by conversion of the alcohol to the corresponding halide. Finally, the halide intermediate was dimerized and dehalogenated in a single step in liquid ammonia with either sodium or potassium amide. This final step was not very efficient, however, generating the pentaene in only 4.8% and the heptaene in only trace amounts [31]. Although both products were isolated as crystalline species, melting points could not be determined due to their rapid decomposition upon heating. The IR spectrum of the pentaene was determined and the UV-visible spectrum was measured for both polyenes (Table 4.6) [31].

It was almost another decade before any further advancements were made. Thus, it was in 1961 that Franz Sondheimer (1926–1981) and his coworkers reported synthetic methods that significantly extended the family to as high as the decaene [32]. As shown in Fig. 4.9, these methods utilized a potassium t-butoxide isomerization method for the generation of the polyenes from various ene-yne precursors. The methods were first utilized to generate the known pentaene from both *trans*-5-decene-1,9-diyne and the analogous allene. In both cases, the desired pentaene was isolated in 9% yield, nearly double that previously reported by Mebane [31].

Similar methods were then utilized to isolate the previously unknown higher polyenes ($n = 6, 8, 10$), with yields for the final isomerization step of ca. 10, 2.5%, and trace amounts, respectively. All of the polyenes were crystalline materials, but as with the previous reports, they rapidly decomposed via polymerization and thus melting points could not be determined. IR spectra were recorded for all but the decaene, while UV-visible absorption data were collected for all species (Table 4.6) [32].

All members of the parent polyene series exhibit the same type of spectra with four main maxima (Table 4.6, A–D), of which the two highest wavelength bands (A, B) are the most intense. In each case, the maxima are spaced 1450 ($\pm$150) cm^{-1}

Table 4.6 UV-visible data for the polyene series, H–(CH=CH)$_n$–H (n = 3–8, 10) [30–32]

n	Band A		Band B		Band C		Band D	
	λ_{max} (nm)	ε ($\times 10^{-3}$)	λ_{max} (nm)	ε ($\times 10^{-3}$)	λ_{max} (nm)	ε ($\times 10^{-3}$)	λ_{max} (nm)	ε ($\times 10^{-3}$)
3[a]	268	34.6	257	42.7	248	30.5	240	19.2
4[b]	304		290		278		267	
5[a]	334	121	317	115	303	71.2	290	37.1
6[a]	364	138	344	127	328	73.2	313	37.3
7[c]	390	36.0[d]	368	33.0[d]	350	20.0[d]	332	9.5[d]
8[a]	410	108[e]	386	112[e]	367	72.8[e]	349	35.8[e]
10[a]	447		420		397		376	

[a]In isooctane, Ref. [32]; [b]In cyclohexane, Ref. [30]; [c]In isooctane, Ref. [31]; [d]This compound was impure, thus giving very low ε values; [e]The lower ε values are thought to be due to some decomposition during the determination of mass [32]

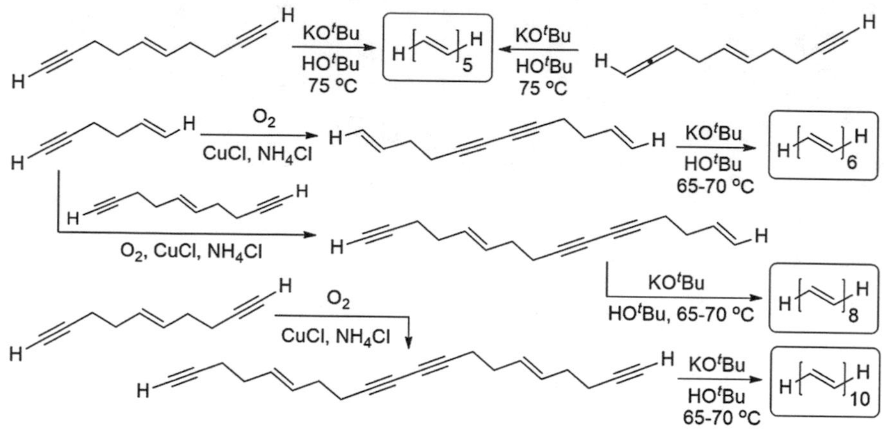

Fig. 4.9 Synthesis of higher unfunctionalized polyenes

apart, typical of conjugated polyenes. Plots of the squares of the absorption energies verses the number of double bonds (n) give straight lines as far as $n = 7$, with small deviations for $n = 8$ and $n = 10$ [32].

4.2 Polyvinylenes

As illustrated by the discussion of polyenes above, many of these species were obtained via the condensation of either water or dihalides. As such, it was only a matter of time before this approach was then applied to polymeric species in an attempt to generate what could be viewed as macromolecular polyenes. The products of these efforts were then referred to as *polyvinylenes* in order to differentiate them from true polyenes.

4.2.1 Polyvinylenes from Poly(vinyl chloride)

The tendency of halogen-containing polymers to lose portions of their halogen content through the catalytic action of light, heat, or chemical reagents had been recognized by the 1930s [33] and thus it is not surprising that early efforts focused on poly(vinyl chloride). As early as 1938, Raymond Fuoss (1905–1987) at General Electric found that dry poly(vinyl chloride) liberated HCl upon heating and that this resulted in an increase in conductivity [34].

The following year, Carl S. Marvel (1894–1988) and coworkers reported that the treatment of poly(vinyl chloride) with zinc in dioxane resulted in smooth dehalogenation with the removal of 84–87% of the chlorine [35]. The resulting material was still soluble in dioxane and further characterization led to the conclusion that the polymer product consisted of cyclopropane units with remaining chlorinated defects (Fig. 4.10). In contrast, treatment of the polymer with KOH was found to remove HCl to give an insoluble, shiny, red-brown polymer which was believed to correspond to "a very long chain polyene" [35]. The proposed structure is given in Fig. 4.10, but this polyvinylene was not studied in any detail and it is unclear if it was thought that this suffered from the same types of defects as the cyclopropane derivative.

Following the successful production of polyacetylene by Natta in 1955,[4] interest in the conjugated products via dehydrohalogenation increased significantly. In 1959, D. E. Winkler proposed a radical mechanism in an attempt to explain the known release of HCl and darkening of poly(vinyl chloride) when exposed to heat or light [36]. This process was proposed to be initiated by radical species generated via oxidation of the polymer when either heated above 180 °C or exposed to UV light. Abstraction of a methylene hydrogen by the radical initiator would then result in the release of a neighboring chlorine radical to stabilize the material via the formation a double bond (Fig. 4.11). The free chlorine atom could then abstract another hydrogen to form HCl and induce the release of a new chlorine radical. Winkler proposed that probability would favor abstraction of a hydrogen from one of the nearest methylene groups, with the removal of an allylic hydrogen the most highly favored. This would then result in a chain reaction growing the conjugation linearly down the polymer backbone in what was sometimes referred to as a "zipper" process. This process could then be terminated by a number of different possible reactions, including reaction of the propagating radical with either oxygen or another radical species.

The following year, Charles Sadron (1902–1993) and coworkers reported the elimination of HCl from poly(vinyl chloride) using quite mild conditions [37]. Via their methods, they utilized various metal salts in the presence of dimethylformamide (DMF) at 80 °C, of which LiCl and LiBr gave the best results. If the solvent used was either pure DMF or combinations of DMF with tetrahydrofuran or dioxane, the polymer solution transitioned to pink and then violet to purple. Alternatively, if DMF/cyclohexanone mixtures were used, the solution transitioned through yellow to orange to brown and finally to black. Both products were characterized via UV-visible spectroscopy, for which the purple solution gave a single transition with a λ_{max} at 530 nm [37]. In contrast, the brown solution exhibited multiple transitions and was concluded to be a mixture of various units of shorter conjugation lengths consistent

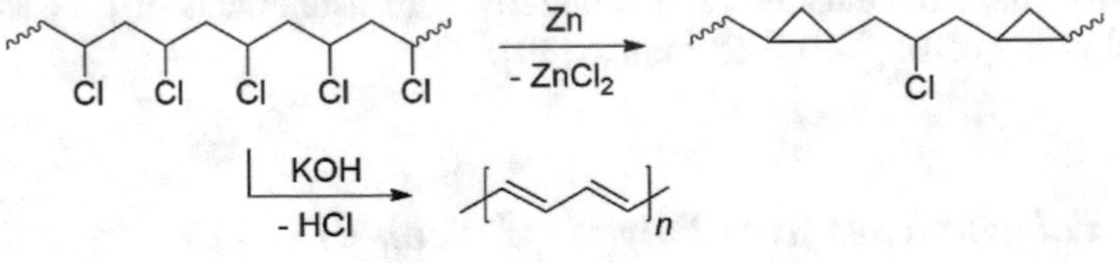

Fig. 4.10 Marvel's dehalogenation of poly(vinyl chloride)

---

[4]See Chap. 5.

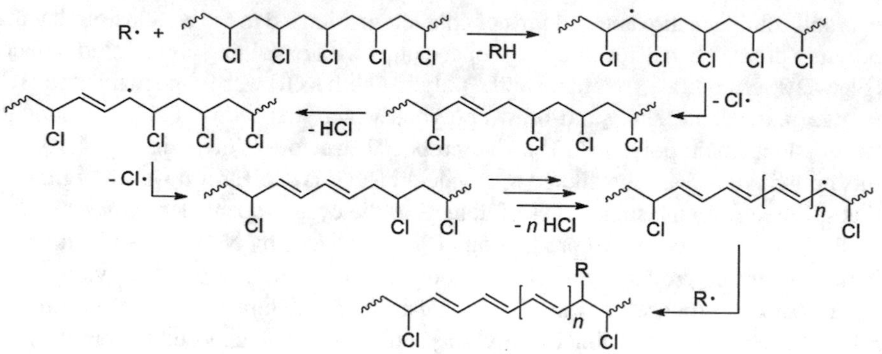

Fig. 4.11 Proposed mechanism for the formation of polyvinylene via radical processes

with the previous spectra of dimethylpolyenes [26]. The authors ultimately concluded that the purple material consisted predominately of conjugated double bonds [37].

The thermal dehydrohalogenation was then studied in more detail by Bengough and Sharpe in 1963 [38]. Heating solutions of the polymer revealed liberation of HCl at temperatures as low as 178 °C. The rate of HCl loss was found to be dependent on both the solvent used and the polymer concentration, with higher rates found for more concentrated solutions. In addition, it was also determined that the rate was dependent on the polymer degree of polymerization, with lower molecular weight materials giving higher rates for HCl evolution. This final dependence was explained by the fact that poly(vinyl chloride) typically contains an alkene endgroup and the chloride adjacent to the double bond should be more reactive than those in a completely saturated environment. As lower molecular weight materials would have a greater percentage of alkene units, this would enhance the rate of HCl evolution [38].

The following year, Tsuchida et al. [39] studied the dehydrohalogenation of poly(vinyl chloride) using sodium amide (NaNH$_2$), investigating the effect of temperature, reaction time, and equivalents of NaNH$_2$. Analysis of the resulting black product found no nitrogen had been incorporated and the ratio of carbon to hydrogen was near unity. The IR spectra exhibited bands at 3030, 1600, and 990 cm^{-1}, which were assigned to the =C–H stretch, C=C stretch, and =C–H bend, respectively. Analysis by X-ray diffraction revealed no evidence of crystallinity. Finally, the electronic properties were characterized by EPR to give results consistent with delocalized, unpaired electrons and conductivity measurements of pressed pellets revealed values of 10^{-3}–10^{-4} Ω^{-1} cm^{-1} [39].

4.2.2 Polyvinylenes from Poly(vinyl alcohol)

In addition to the dehydrohalogenation of poly(vinyl chloride), various authors investigated the generation of polyvinylenes via the dehydration of poly(vinyl alcohol) beginning in the early 1960s. These efforts began in early 1960 with a study by Kaesche-Krischer and Heinrich on the thermal degradation of poly(vinyl alcohol) [40]. Heating the material under vacuum (ca. 10^{-2} Torr) while measuring any change in mass revealed that under 100 °C only a small amount of adsorbed water was removed (ca. 1.5% of the total weight). Between 100 and 260 °C, however, a substantial amount of mass was lost, corresponding to 32% of the total. This loss maximized at ca. 160 °C and was accompanied with a deepening of the samples color from white to rusty brown. Isolation of the volatile material lost confirmed that it was water and the color change was thus attributed to the formation of double bonds as a result of the dehydration. It was concluded, however, that this dehydration was not complete, as complete dehydration should theoretically result in a loss of 41% of the total mass. If the temperature was further raised, addition mass was lost between 260 and 350 °C, which was concluded to be various aldehydes as a result of decomposition pathways [40].

Chemical dehydration was then reported by Mortimer M. Labes and coworkers in 1962, who studied the treatment of poly(vinyl alcohol) with fuming H_2SO_4 to generate a black viscous solution [41]. From this solution, a fine amorphous black powder was isolated that was insoluble in all organic solvents. The IR spectra revealed bands consistent with the previous dehydrohalogenation products of poly(vinyl chloride) and the resistivity of the material as a pressed pellet was determined to be 10^7 Ω cm. It was concluded that even after extensive drying this material was still regarded as quite impure.

The thermal decomposition of poly(vinyl alcohol) was then reported again by Tsuchiya and Sumi in 1969 [42]. They determined dehydration to begin at ca. 200 °C, quite higher temperatures than that previously reported. Heating the material at 240 °C for 4 h, it was determined that 33.4% of the weight was lost as water, along with very small amounts of various other organic species (primarily aldehydes and ketones). The determined extent of dehydration was in excellent agreement with the previous work of Kaesche-Krischer and Heinrich [40]. Further heating at 450 °C then resulted in the loss of additional material (ca. 27.7% of the total mass), which was determined to be primarily various organic species.

References

1. Rasmussen SC (2014) The path to conducting polyacetylene. Bull Hist Chem 39:64–72
2. Rasmussen SC (2017) Cuprene: a historical curiosity along the path to polyacetylene. Bull Hist Chem 42:63–78
3. Shirakawa H (2001) The discovery of polyacetylene film: the dawning of an era of conducting polymers. In: Frängsmyr T (ed) Les Prix Nobel. The Nobel Prizes 2000, Nobel Foundation, Stockholm, pp 217–226
4. Shirakawa H (2001) The discovery of polyacetylene film: the dawning of an era of conducting polymers (Nobel lecture). Angew Chem Int Ed 40:2574–2580
5. Shirakawa H (2001) Nobel lecture: the discovery of polyacetylene film—the dawning of an era of conducting polymers. Rev Mod Phys 73:713–718

6. Shirakawa H (2002) The discovery of polyacetylene film. The dawning of an era of conducting polymers. Synth Met 125:3–10

7. Kuhn R, Winterstein A (1928) Über konjugierte Doppelbindungen I. Synthese von Diphenyl-poly-enen. Helv Chim Acta 11:87–116

8. Kuhn R, Winterstein A (1928) Über konjugierte Doppelbindungen II. Synthese von Biphenylen-poly-enen. Helv Chim Acta 11:116–122

9. Kuhn R, Winterstein A (1928) Über konjugierte Doppelbindungen III. Wasserstoff- und Brom-anlagerung an Poly-ene. Helv Chim Acta 11:123–144

10. Kuhn R, Winterstein A (1928) Über konjugierte Doppelbindungen IV. Molekelverblndungen und Farbreaktionen der Poly-ene. Helv Chim Acta 11:144–151

11. Baer HH (1969) Richard Kuhn 1900–1967. Adv Carbohydr Chem 24:1–11

12. Baer HH (1993) Richard Kuhn (1900–1967): Werk und Persönlichkeit. Liebigs Ann Chem 1993(11):I–XXX1

13. Kuhn R (1937) Über die Synthese höherer Polyene. Angew Chem 50:703–708

14. Hengstenberg J, Kuhn R (1930) Die Kristallstruktur der Diphenylpolyene. Z Kristallogr Kristallgeom Kristallphys Kristallchem 75:301–310

15. Hausser KW, Kuhn R, Smakula A (1935) Lichtabsorption und Doppelbindung. IV. Diphenylpolyene. Z Phys Chem Abt B 29:384–389

16. Hausser KW, Kuhn R, Seitz G (1935) Lichtabsorption und Doppelbindung. V. Über die Absorption von Verbindungen mit konjugierten Kohlenstoffdoppelbindungen bei tiefer Temperatur. Z Phys Chem Abt B 29:391–416

17. Hausser KW, Kuhn R, Kuhn E (1935) Lichtabsorption und Doppelbindung. VI. Über die Fluorescenz der Diphenylpolyene. Z Phys Chem Abt B 29:417–454

18. Kuhn R, Grundmann C (1938) Uber die Synthese von 1.6-Dimethyl-hexatrien, 1.8-Dimethyl-octatetraen und 1.12-Dimethyl-dodecahexaen. Ber Dtsch Chem Ges 71B:442–447

19. Blout ER, Fields M (1948) Absorption spectra. V. The ultraviolet and visible spectra of certain polyene aldehydes and polyene azines. J Am Chem Soc 70:189–193

20. Blout ER, Fields M, Karplus R (1948) Absorption spectra. VI. The infrared spectra of certain compounds containing conjugated double bonds. J Am Chem Soc 70:194–198

21. Blout ER (2002) Oral History Interview conducted by Bohning JJ, Thackray A (Transcript #0263) at Harvard Medical School, Harvard School of Public Health, and Cambridge, Massachusetts. Chemical Heritage Foundation, Philadelphia, 30 May 1991, 13 September and 22 November 2002

22. Pearce J (2007) Elkan R. Blout, Scientist at Harvard, Dies at 86. The New York Times, January 9

23. Bohlmann F (1952) Reaktionen mit Lithiumaluminiunhydrid: Reduktion von Polyen-1.2-diketonen, zugleich eine Methode zur Darstellung von Polyenen. Chem Ber 85:386–389

24. Winterfeldt E (1994) Ferdinand Bohlmann (1921–1991) und sein wissenschaftliches Werk. Liebigs Ann Chem 1994(5):I–XXXIV

25. Bohlmann F (1946) Solvatochromie in der Pyridinreihe. Ph.D. Dissertation, University of Göttingen

26. Karrer P, Coehand C (1945) Di-[ω-phenyl-polyen]-diketone. Helv Chim Acta 28:1181–1184

27. Bohlmann F, Mannhardt HJ (1956) Konstitution und Lichtabsorption, VIII. Mitteil. 1): Darstellung und Lichtabsorption von Dimethyl-polyenen. Chem Ber 89:1307–1315

28. Wittig G, Schöllkopf U (1954) Über Triphenyl-phosphin-methylene als olefinbildende Reagenzien (I. Mitteil.). Chem Ber 87:1318–1330

29. Woods GF, Schwartzman LH (1948) 1,3,5-Hexatriene. J Am Chem Soc 70:3394–3396

30. Woods GF, Schwartzman LH (1949) 1,3,5,7-Octatetraene. J Am Chem Soc 71:1396–1399

31. Mebane AD (1952) 1,3,5,7,9-Decapentaene and 1,3,5,7,9,11,13-tetradecaheptaene. J Am Chem Soc 74:5527–5529

32. Sondheimer F, Ben-Efraim DA, Wolovsky R (1961) Unsaturated macrocyclic compounds. XVII. The prototropic rearrangement of linear 1,5-enynes to conjugated polyenes. The synthesis of a series of vinylogs of butadiene. J Am Chem Soc 83:1675–1681

33. Boyer RF (1947) A statistical theory of discoloration for halogen-containing polymers and copolymers. J Phys Colloid Chem 51:80–160
34. Fuoss RM (1938) Preparation of polyvinyl chloride plastics for electrical measurements. Trans Electrochem Soc 74:91–112
35. Marvel CS, Sample JH, Roy MF (1939) The structure of vinyl polymers. VI. Polyvinyl halides. J Am Chem Soc 61:3241–3244
36. Winkler DE (1959) Mechanism of poly(vinyl chloride) degradation and stabilization. J Polym Sci 35:3–16
37. Sadron C, Parrod J, Roth JP (1960) On the dehydrochlorination of polyvinyl chloride. Compt Rend 250:2206–2208
38. Bengough WI, Sharpe HM (1963) The thermal degradation of polyvinylchloride in solution. I. The kinetics of the dehydrochlorination reactin. Makromol Chem 66:31–44
39. Tsuchida E, Shih CN, Shinohara I, Kambara S (1964) Synthesis of a polymer chain having conjugated unsaturated bonds by dehydrohalogenation of polyhalogen-containing polymers. J Polym Sci A 2:3347–3354
40. Kaesche-Krischer B, Heinrich HJ (1960) Pyrolyse und Entzündung von Polyvinylalkohol (PVA). Z Phys Chem 23:292–296
41. Mainthia SB, Kronick PL, Labes MM (1962) Electrical measurements on polyvinylene and polyphenylene. J Chem Phys 37:2509–2510
42. Tsuchiya Y, Sumi K (1969) Thermal decomposition products of poly(vinyl alcohol). J Polym Sci A: Polym Chem 7:3151–3158

Chapter 5
Polyacetylene

Although the polymerization of acetylene had been studied since nearly its first isolation and large-scale production, none of the materials produced via these early attempts corresponded to the highly conjugated material that is now known as polyacetylene [1, 2]. The production and study of the material most commonly associated with acetylene polymers had to wait until the development of metal-based polymerization catalysts by Karl Ziegler (1898–1973)[1] in the early 1950s [3–5]. This then led to the application of these catalysts to the polymerization of first olefins and then acetylenes by Giulio Natta (1903–1979) of Milan, Italy.

5.1 Natta and the Polymerization of Acetylenes

Giulio Natta (Fig. 5.1) was born in the small Italian city of Porto Maurizio (now Imperia), near the French border, on February 26, 1903 [6–11]. His father was a judge in Genoa [7, 10], where the family spent the winters [10]. It was also here that Giulio attended the 'Christopher Columbus' school [7, 11], before continuing on to the University of Genoa to study mathematics [7, 10]. After two years, however,

---

[1] Karl Ziegler was born on November 26, 1898 in Helse near Kassel in Germany [3, 4]. He studied at Marburg University, taking his doctorate under von Auwers in 1920 [3, 4], after which he qualified as a lecturer in 1923 [4]. He then moved to the University of Frankfurt in 1925, but moved again to become professor at the University of Heidelberg in 1927 [3]. He moved to become professor and director of the Chemisches Institut at the University of Halle in 1936, after which he was offered the directorship of the Kaiser-Wilhelm-Institut fur Kohlenforschung at Mülheim in 1943 [3, 4]. He was happy at Halle, however, and only moved to Mülheim after the end of World War II when the Americans occupying Halle in the Spring of 1945 urged him to move before the Russians took over the region under the Yalta agreement [3]. Best known for his development of a new class of polymerization catalysts and their application to high quality, linear polyethylene, Ziegler shared the Nobel Prize in Chemistry with Natta in 1963 [3, 4]. Ziegler retired as director of the Mulheim Institute in 1969, but stayed on as an honorary research fellow. He died on August 11, 1973 [3].

© The Author(s) 2018
S. C. Rasmussen, *Acetylene and Its Polymers*, SpringerBriefs in Molecular Science,
https://doi.org/10.1007/978-3-319-95489-9_5

Fig. 5.1 Giulio Natta
(1903–1979) (Courtesy of
the Giulio Natta Archive)

he changed disciplines and moved to the Polytechnic School of Milan in 1921 [9, 10] to study chemical engineering [6, 7, 10, 11]. In 1922, he began research under Giuseppe Bruni (1873–1946)[2], Director of the Institute of General Chemistry, and Giorgio Renato Levi (1895–1965) [10]. In 1924, Natta then received his *Dottore* degree in chemical engineering [6–11] and continued on as Bruni's assistant [8, 10]. He attained the position of *Libero docente* in 1927 [6, 7], which then allowed him to teach [6]. During a 1932 visit to Freiburg to study the work of Hugo Seeman on the new technique of electron diffraction, Natta met Hermann Staudinger (1881–1965), who suggested that he use X-ray crystallography to study polymer structures [8].

After a period as an assistant lecturer in chemistry at Milan [7], he moved to the University of Pavia in 1933, where he was appointed full professor and director of the Institute of General Chemistry [6–9, 11]. After only two years, however, he then moved to the University of Rome to occupy the chair of physical chemistry in 1935 [6–9, 11]. The same year he married Rosita Bead, a graduate of the arts [10]. As with Pavia, he didn't stay at Rome long, leaving to take the chair of industrial chemistry at the Polytechnic of Turin in either 1936 [6] or 1937 [7, 8, 11]. Finally, he was called back to his alma mater to become the chair of industrial chemistry of the Polytechnic of Milan in either 1938 [6–9, 11] or 1939 [10]. It was here that Natta's focus turned to polymers [8] and where he remained until his retirement in 1973 [7, 10, 11]. Natta began to suffer from Parkinson's disease in 1959 [7–10], severely

---

[2]Giuseppe Bruni was born in Parma, Italy in 1873 [12]. After graduating from Parma in 1896, he joined Ciamician at Bologna, where he focused on the nature of solid solutions and isomorphism. During the period 1900–1901, he traveled to work with van't Hoff before returning to Bologna. He was then appointed to the Chair of General and Inorganic Chemistry at Padua in 1907, where he remained until 1917 when he was appointed professor of general and inorganic chemistry at the Polytechnic School of Milan. He held this position until shortly before his death in 1949 [12].

restricting his mobility, although he insisted on continuing his research as much as possible [8, 10]. He died only six years after his retirement, on May 1, 1979 at the age of 76 [7, 8, 10].

Natta is best known for his work in high molecular weight polymers, beginning upon his return to Milan with work on butadiene and synthetic rubber in 1938 [6–9]. That same year he also began work on the polymerization of olefins [6, 7], which ultimately led to the extension of Karl Ziegler's work on metal-based polymerization catalysts in the early 1950s. These efforts then resulted in the discovery of new classes of polymers with a sterically-ordered structure (i.e. isotactic, syndiotactic and di-isotactic polymers), as well as linear non-branched olefinic polymers and copolymers with an atactic structure (Fig. 5.2) [6–11]. For these accomplishments, Natta shared the Nobel Prize in Chemistry with Ziegler in 1963 [7–11].

After his various successes in the application of catalytic polymerization to α-olefins and diolefins in the early 1950s, Natta expanded his scope and began investigating the application of the previously successful catalysts to the polymerization of acetylenes [13–15]. These efforts were motivated not only as an extension of his previous work with olefins, but also from the fact that it was already known that acetylene polymerizations could be catalyzed by metal species, as previously reported for the generation of cuprene [1, 13, 15]. These efforts resulted in an initial Italian patent granted in July of 1955 [13], which encompassed the polymerization of acetylene and its alkyl or aryl derivatives using organometallic catalysts of group 4-8 transition metals. Specific examples of catalytic systems included titanium species (titanium

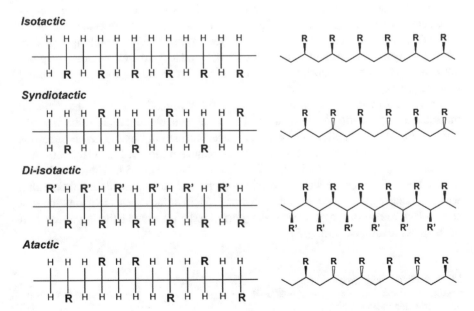

Fig. 5.2 Fischer projections and line drawings of various sterically-ordered structural types of polyolefins

trichloride, titanium tetrachloride, or titanium alkoxides) with triethylaluminum or diethylaluminum chloride [13].

Natta then presented the results of some of his initial acetylene polymerizations in July of 1957 at the XVI International Congress for Pure and Applied Chemistry in Paris, a summary of which was included as part of a report of the meeting published in *Angewandte Chemie* later that same year [14]. Using catalyst combinations of titanium halides or alkoxides with alkylaluminum species, they reported the successful polymerization of acetylene and its methyl, ethyl, and butyl derivatives. Although the parent acetylene gave a solid material that was insoluble in organic solvents, the alkyl derivatives gave rubber-like polymers soluble in organic solvents. Of these, the product of the polymerization of 1-hexyne was then studied in greater detail in order to draw conclusions about the polymer structure [14].

The polymer was first fractionated to isolate the portion that was insoluble in acetone, but soluble in ether, and it was this fraction that was then used in the sequential investigations. The polymer was first subjected to colorimetric wet chemical tests for double bonds, the results of which were consistent with the presence of double bonds as expected. Although it was found to be difficult to hydrogenate the polymer product using Raney nickel, the infrared (IR) spectrum of the hydrogenated product was very similar to that of the polyhexane polymerized under similar conditions. In contrast, the polymer product was easily oxidized with perbenzoic acid, consuming ca. 0.85 equivalents of perbenzoic acid per repeat unit to give what was believed to be corresponding peroxide derivative. The initial oxidation product was then hydrolyzed and further oxidized with lead(IV) acetate. Analysis of this oxidative degradation revealed products of valeric acid along with smaller amounts of butyric and propionic acid. Lastly, the IR spectrum of the polymer product revealed peaks consistent with butyl groups bonded to unsaturated carbon atoms, which are no longer present after hydrogenation. The bulk of these results then led to the conclusion that the product was polyhexyne, "a substantially linear, highly unsaturated polymer containing C_4 side groups" [14].

This initial report was then followed by a full publication in 1958 that detailed the successful catalytic polymerization of acetylene via combinations of triethylaluminum (Et_3Al) and titanium alkoxides [15]. As outlined in Fig. 5.3, the best results were obtained by bubbling acetylene into a heptane solution of Et_3Al (0.1 mol) and titanium(IV) propoxide (0.04 mol) at 75 °C over a period of 15 h. These conditions resulted in a 98.5% conversion of acetylene to give a black, crystalline polymer that was completely insoluble in organic solvents [15]. In some cases, the formation of a shiny black mirror of the polymer was reported to be observed adhering to the wall of the glass reaction vessels [15].

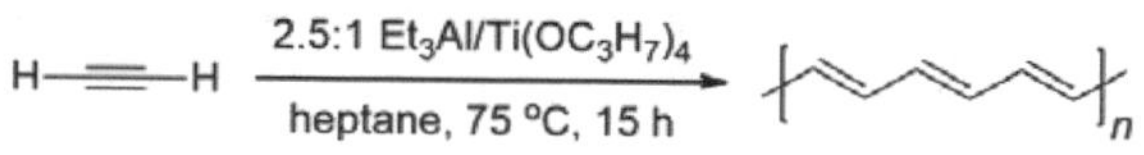

Fig. 5.3 Natta's optimized conditions for the polymerization of acetylene

The powder samples of the polymer products were characterized by X-ray diffraction (CuKα), which were found to be ~90–95% crystalline, with low amorphous content. Lattice spacings of $d_1 = 3.65$ Å and $d_2 = 2.11$ Å were found, which were noted to be related by the expression $d_1 = \sqrt{3}d_2$. This suggested that these reflections came from lattice planes parallel to the axes of linear polymer chains of a small cylindrical enclosure, densely packed in a pseudohexagonal lattice at relative distances of 4.22Å $= (2/\sqrt{3})d_1 = 2d_2$ (Fig. 5.4) [15]. Overall, it was determined that the collected X-ray data were consistent with linear chains of polyacetylene in which the double bonds were concluded to be predominantly *trans* in configuration.

The combination of the material's black color, metallic luster, and relatively low electrical resistivity (ca. 10^{10} Ω cm, compared to 10^{15}–10^{18} Ω cm for typical polyhydrocarbons) led Natta to conclude that these polyacetylene products consisted of long sequences of conjugated double bonds. As such, these polyacetylene materials were structurally identical to a very long polyene.[3] Hideki Shirakawa[4] later stated, however, that this conclusion was not accepted widely at the time [16–19].

Lastly, some limited study of the polymer's chemical properties was also attempted. The complete insolubility of the material prohibited any possibility to determine the molecular weight and attempts to use solvents to fractionate the material were unsuccessful. Even treatments with solvent at high temperature failed to dissolve the polymer, only resulting in a reduction in crystallinity. For example, after treatment with boiling tetralin (1,2,3,4-tetrahydronaphthalene; bp 206–208 °C) under a nitrogen atmosphere, the crystallinity of the polymer was reduced to only about 25%. This loss of crystallinity was attributed to crosslinking [15].

It was also noted that the polymer samples exhibited higher reactivity than comparable polyolefins, particularly with oxidants such as O_2 and Cl_2. For example, the materials rapidly absorbed atmospheric oxygen at high temperatures, resulting in the formation of more lightly colored products. Reaction with chlorine occurred even at room temperature, resulted in a white solid that was amorphous by X-ray char-

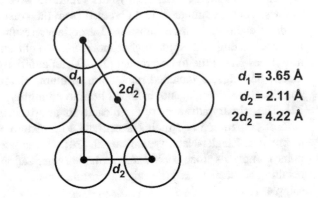

Fig. 5.4 Proposed polyacetylene packing from Natta's X-ray diffraction data

$d_1 = 3.65$ Å
$d_2 = 2.11$ Å
$2d_2 = 4.22$ Å

---

[3] See Chap. 4.
[4] See Sect. 5.5.

acterization. Heating this product at 70–80 °C resulted in a rapid darkening of the polymer and loss of HCl. Alternately, nearly all of the chlorine could be removed by treatment of the chlorinated polymer with potassium in hot ethanol, giving a black amorphous powder. Natta pointed out that this high reactivity with chlorine clearly differentiated his polyacetylene from cuprene [15].

Although Natta states in this final 1958 paper that these results only represented an initial communication with additional reports planned (particularly a more detailed report on his polyhexyne results) [15], no further work on polyacetylenes was ever published by Natta and his coworkers. Others, however, did not hesitate to continue what Natta had started. As a result, the formal name polyacetylene gradually replaced the term polyene as more studies began to utilize Natta's polymerization methods [16–19] and the older term became strictly associated with oligomeric species.

5.2 Additional Studies of Acetylene Polymerizations

Much of the initial extension of Natta's work on polyacetylene focused not only on the further study of the polymerization reaction and the materials generated, but on new catalytic systems for the polymerization of acetylene. This work was largely led by L. B. Luttinger of the American Cyanamid Company beginning in 1960 [20–22]. The system reported by Luttinger consisted of a mixture of a hydride reducing agent with a metal salt or complex from groups 8–10. Typical combinations consisted of sodium borohydride with either nickel halide (or an analogous complex such as $(Bu_3P)_2NiCl_2$) or cobalt nitrate [20, 21]. These catalytic mixtures allowed the smooth and rapid polymerization of acetylene or mono-substituted derivatives even at room temperature. Application of the $NaBH_4/Co(NO_3)_2$ mixture to the polymerization of acetylene produced a black solid material whose IR spectrum and X-ray diffraction pattern was consistent with the previous results of Natta [15, 20]. Application of the $NaBH_4/(Bu_3P)_2NiCl_2$ mixture gave more variable results, with the best samples similar to that obtained from the cobalt system [21]. Application of these catalysts to monosubstituted acetylenes resulted in primarily oligomeric materials, along with some higher molecular weight material, while disubstituted acetylenes gave little to no product [20, 21]. In addition to nickel and cobalt, other metals investigated included ruthenium, osmium, platinum, and palladium, while additional hydride reagents included lithium aluminium hydride and diborane [20, 22]. The primary advances of these catalysts as advocated by Luttinger were their ease of preparation and their insensitivity to moisture and oxygen in comparison to previous Ziegler-Natta type catalysts [20]. It was later shown that nickel phosphine complexes alone could catalyze the polymerization of acetylene, although the resulting products lacked the shiny appearance characteristic of other polyacetylene samples [23].

Shortly after Luttinger introduced his new catalyst system, John K. Stille (1930–1989)[5] and D. A. Frey of the University of Iowa reported the extension of Natta's methods to the polymerization of non-conjugated diynes such as 1,6-heptadiyne [24]. The polymerization of 1,6-heptadiyne gave dark red to black products, indicative of a significant amount of conjugation within the product. The black products were produced using the previously optimized conditions of Natta and were found to be completely insoluble. In contrast, the use of titanium chlorides in larger molar quantities resulted in red, soluble products. The analysis of one of the soluble materials gave a number-average molecular weight (M_n) of 13,500. Overall characterization of the various samples gave very similar properties and reactivity to that reported by Natta for the unfunctionalized parent [15] and a combination of reductive and oxidative decomposition studies led to the proposed structure given in Fig. 5.5. Attempts to polymerize 1,7-octadiyne and 1,8-nonadiyne gave only lightly colored products in low yields [24].

Later that same year, William H. Watson of Texan Christian University revisited the polymerization of acetylene via Zieger-Natta catalysts in an attempt to provide a more detailed study of the process [27]. In this report, he cited the rising interest in organic semiconductors and the emphasis being placed upon the electronic properties of long chain polyenes as his motivation. In Watson's study, the transition metal applied was titanium(IV) chloride (in comparison to Natta's titanium(IV) propoxide [15]), coupled with a variety of metal alkyls, including tri(isobutyl)aluminum ($^{i}Bu_3Al$), butyl lithium, and diethylzinc [27]. Although the most effective catalyst combinations and conditions were in close agreement with Natta's previous results (Fig. 5.6), Watson did provide much more detailed descriptions of the experimental conditions of the polymerization process. Such details included the need to use cooling baths to maintain a constant reaction temperature, the flow rate of the added acetylene gas (350 mL min^{-1}), and the fact that extensive washing of the product was necessary in order to remove large quantities of inorganic salts trapped in the polymer. Surprisingly, however, Watson never seems to specify the reaction time employed.

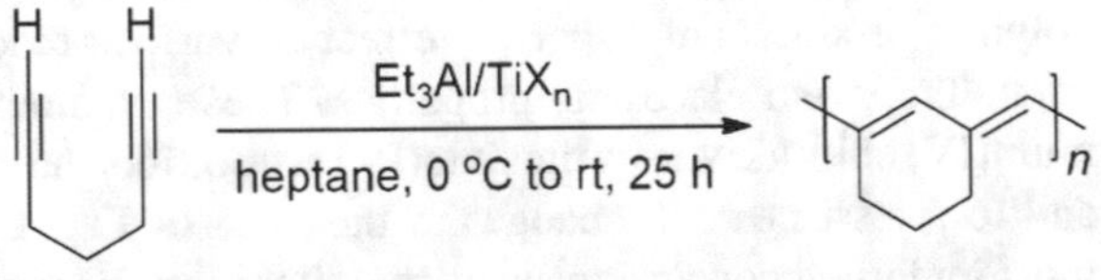

Fig. 5.5 Polymerization conditions and proposed structure for poly(1,6-hexadiyne)

Et_3Al/TiX_n

heptane, 0 °C to rt, 25 h

[5]John K. Stille was born in Tucson, Arizona, on May 8, 1930 [25, 26]. He received his B.A. in Chemistry at the University of Arizona in 1952, followed by his M.A. the following year [25, 26]. He then served in the Navy during the Korean War [25] before earning his Ph.D. at the University of Illinois in 1957 under Carl Marvel (1894–1988) [25, 26]. He joined the faculty at the University of Iowa in 1957 [25, 26], becoming professor in 1965 [26]. He later moved to Colorado State University in 1977, where he was appointed university distinguished professor in 1986 [26]. He made important contributions to both of the fields of polymer chemistry and organometallic chemistry, but is best known for the cross-coupling reaction that bears his name [26]. He died on July 19, 1989, in the crash of a United Airlines flight in Sioux City, Iowa [25].

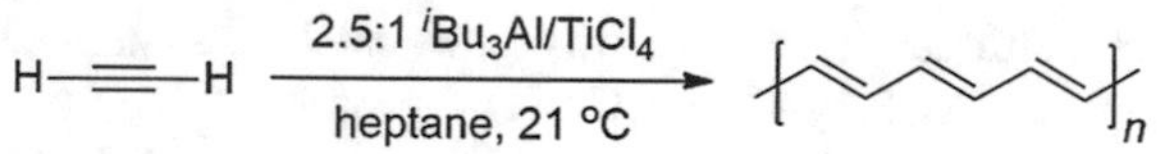

$$H\!-\!\!\equiv\!\!-\!H \xrightarrow[\text{heptane, 21 °C}]{2.5:1\ ^i\text{Bu}_3\text{Al/TiCl}_4} \left\{\!\!\diagdown\!\!\diagup\!\!\diagdown\!\!\diagup\!\!\diagdown\!\!\right\}_n$$

Fig. 5.6 Watson's optimized conditions for the polymerization of acetylene

The polymers generated were insoluble black materials that did not melt below 400 °C and analysis indicated an empirical formula of $(CH)_n$. Some limited characterization of the electronic properties was also reported, including room temperature paramagnetic resonance measurements to give values of 0.94–14.6×10^{18} spin g^{-1} and electrical measurements to give a resistance of ca. 10^4 Ω cm at 25 °C (considerably lower than that reported by Natta, ca. 10^{10} Ω cm [15]). Lastly, the electrical measurements exhibited a temperature dependence typical of semiconductors [27].

Although Watson stated that more detailed reports of the electronic properties of these polymeric products would follow, he never published any further reports on acetylene polymerizations. This may have been because of a paper published shortly thereafter by Masahiro Hatano (b. 1930) at the Tokyo Institute of Technology, which reported the exact electronic characterization alluded to by Watson.

5.3 Masahiro Hatano

Masahiro Hatano was born in 1930 and received his doctorate from the Tokyo Institute of Technology in 1959. He then joined the Chemical Resources Laboratory at Tokyo Institute of Technology and remained there until 1967, when he moved to the Chemical Research Institute of Non-Aqueous Solution at Tohoku University as an associate professor. He became professor in 1969 and professor emeritus in 1994.

Shortly after receiving his doctorate, Hatano reported the first detailed study characterizing the semiconducting properties of polyacetylene in 1961 [28]. The study began with an investigation of the Ziegler-Natta catalyzed polymerization conditions. Unlike previous studies, however, the focus was not on increasing the yield of solid polymer product, but rather the effect of various reaction parameters on polymer crystallinity and electronic properties. These parameters included the use of titanium(IV) chloride versus titanium(IV) propoxide, the effect of reaction temperature, and to a lesser extent, changes in the ratio of Et_3Al to the titanium species. The most important conclusions were that titanium(IV) propoxide gave more crystalline materials than the corresponding chloride and that crystallinity generally increased with the reaction temperature [28].

Hatano then characterized the polymer products via electron spin resonance (ESR) spectroscopy and pressed-pellet DC conductivity measurements, with particular emphasis on the effect of crystallinity [28]. As illustrated by the selected results summarized in Table 5.1, samples of increased crystalline nature exhibited a greater concentration of unpaired electrons and lower resistance. The values for the crystalline samples here are in relatively good agreement with the previous results

Table 5.1 Selected electronic properties of polyacetylenes of variable crystallinity [28]

Polymer crystallinity	Concentration of unpaired electrons (spin g^{-1})	Resistance (Ω cm)
Amorphous	4.4×10^{18}	3.7×10^9
Low crystallinity	11×10^{18}	1.6×10^8
Medium crystallinity	36×10^{18}	4.2×10^5
High crystallinity	47×10^{18}	1.4×10^4

of Watson [27], while the resistance of the amorphous sample is in good agreement with that previously reported by Natta [15]. Lastly, the resistance was determined over the range of ca. 20–125 °C, which exhibited a temperature dependence typical of intrinsic semiconductors (Fig. 5.7), as previously reported by Watson [27].

Some studies of the effects of air oxidation on electronic properties were also reported [28]. A sample of amorphous polyacetylene underwent oxidization at room temperature over 15 days, resulting in a color change from greenish black to pale orange and an increase in oxygen content as determined by chemical analysis. Characterization of the oxidized polymer sample indicated a decrease in the concentration of unpaired electrons (0.3×10^{18} spin g^{-1}) and an increase in resistivity (1.4×10^{12} Ω cm).

The following year Hatano and coworkers continued these efforts by investigating new types of polymerization catalysts with the goal of producing highly crystalline polyacetylene [29]. Based on the previous study [28], they concluded that the use of transition metal halides generated amorphous polyacetylene samples, while the application of the analogous alkoxides gave crystalline materials. As such, this follow-up study focused on the potential of various transition metal acetylacetonates as active

Fig. 5.7 Resistivity versus 1/T for an amorphous polyacetylene sample (Reproduced from Ref. [28], © 1961 Interscience Publishers, Inc., New York, used with permission)

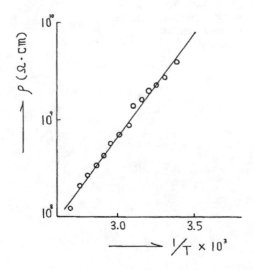

catalysts for the generation of crystalline polyacetylenes [29]. The study investigated catalyst mixtures containing metal complexes of acetylacetone (acac) with Et$_3$Al, in which the metals studied included Ti, V, Cr, Fe, Co and Cu. Of these metal complexes, however, only two species, Ti(acac)$_2$ and VO(acac)$_2$ were effective catalysts and generated crystalline products when combined with Et$_3$Al.

This study was then followed by a collaboration with Hatano's colleague Sakuji Ikeda (1920–1984), who will be covered in more detail in the following section. Thus, the two researchers published a review of organometallic complexes as polymerization catalysts in 1963 [30]. Although this review did include catalytic polymerization of acetylene, most of it focused on the polymerizations of olefins.

Hatano then published two additional reports at the end of his time at the Tokyo Institute of Technology [31, 32]. The first of these reports focused on the studies of the effect of pressure on the conductivity of polyacetylene samples. Here, highly crystalline samples of polyacetylene were produced as previously reported and then pressed into solid 0.7 mm thick pellets. The pellets were then cut and fitted into a high-pressure cell for the conductivity measurements. While holding the cell at a constant pressure, the conductivity was then measured over range of temperatures from room temperature to that of liquid nitrogen [31]. The current-voltage linearity was examined over an applied voltage of 0–50 V, which exhibited an Ohmic relationship, and in all cases the temperature dependence was typical of intrinsic semiconductors as previously observed at atmospheric pressure. As the pressure was increased, however, the overall conductivity also increased (Fig. 5.8). This ultimately led to the conclusion that the charge carrier density changed under pressure, but that the mobility of the charge carriers did not change [31].

Fig. 5.8 Plots of conductivity versus 1/T as a function of applied pressure (Reproduced from Ref. [31], © 1967 The Physical Society of Japan, used with permission)

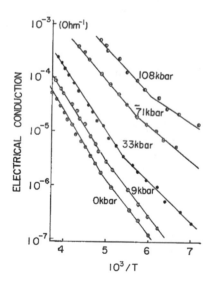

Hatano's final polyacetylene paper was an attempt to use his research to date to present a more well-developed model of the polymer structure, a proposed polymerization mechanism which could account for these structural aspects, and a summary of the known relationships between structure and conductivity [32]. Much of this involved detailed investigations of both polyacetylene and its deuterated analogue via IR spectroscopy, which led to the conclusion that the refractive index of the material was quite high (close to 2), as was its specific gravity. The most significant transition at 1010 cm^{-1} was attributed to an out-of-plane bending vibration of a C–H unit that was part of a *trans* carbon–carbon double bond. However, a transition at 740 cm^{-1} was concluded to correspond to analogous *cis* units and it was found that the relative *cis* and *trans* content in the polymer was dependent on the polymerization temperature [32].

This structural model and the associated temperature dependence then led to the proposed polymerization mechanism given in Fig. 5.9. As proposed, one of the π-bonds of the acetylene was broken as it added to the catalyst to give a *cis*-alkene unit and additional acetylene additions resulted in a trimeric intermediate structure. At this point, the growing end of the chain could either cycle back to eliminate a benzene from the catalyst (process 1), or growth could continue to generate an open chain of double bonds with a *cisoid* conformation (process 2). The *cisoid* conformation could also undergo isomerization to a corresponding *transoid* conformation (process 3), although this process was believed to have a sufficiently large activation energy compared to the other processes and was still believed to be mediated by the polymerization catalyst [32].

Efforts to probe the structure of polyacetylene then continued with X-ray diffraction of samples polymerized at 80 °C, which the previous IR studies had suggested to consist of a nearly pure *trans* conformation. The X-ray diffraction pattern suggested a hexagonal packing and exhibited features similar to that observed for carbon black. As a result, Hatano concluded that polyacetylene had a planar molecular structure and a layered crystal structure [32]. Lastly, X-ray diffraction of polyacetylene samples of variable crystallinity was utilized to develop a more detailed relationship between the polymer crystallinity and resistivity.

After his move to Tohoku University in 1967, Hatano continued to study polymerization processes and polymeric materials, but did not continue any further studies of polyacetylenes. His colleague Sakuji Ikeda, however, continued the tradition of

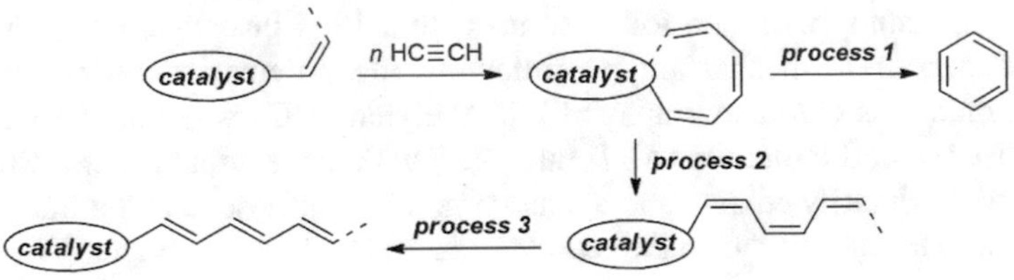

Fig. 5.9 Hatano's proposed polymerization mechanism

polyacetylene studies at Tokyo Institute of Technology begun by Hatano. As such, we will continue with a discussion of Ikeda's research.

5.4   Sakuji Ikeda

Sakuji Ikeda was born in 1920 and graduated from the Department of Applied Chemistry of Tokyo Institute of Technology in 1942. He then received his doctorate from the same institution in 1961, after which he joined the Chemical Resources Laboratory at Tokyo Institute of Technology. He became professor in 1967 and served as the Director of the Chemical Resources Laboratory from 1973 to 1976 [33]. Ikeda then moved to the Nagaoka University of Technology to become the University President and remained so until his death in 1984.

Following his doctorate, Ikeda turned his interest to Ziegler-Natta polymerizations, starting with a review of recent advances in polymerization processes in 1962 [34]. Although the majority of this review focused on olefin polymerizations, section V of the paper covered the polymerizations of alkynes, both acetylene and its functionalized derivatives. This was then followed with a related review of organometallic complexes as polymerization catalysts, co-authored with Hatano in 1963 [30]. Again, this review focused primarily on olefin polymerizations but did include some catalytic polymerizations of acetylene.

Ikeda then utilized the fact that benzene was produced as a byproduct during the Ziegler-Natta polymerization of acetylene to investigate the synthesis of the isotopically labeled species benzene-$^{14}C_6$ and benzene-$^2H_6$ from either acetylene-$^{14}C_2$ or acetylene-$^2H_2$ [35]. The catalyst combination used was Et_3Al and $TiCl_4$ and it was known that the Al/Ti ratio utilized could affect the extent of benzene production relative to polyacetylene. Thus, Ikeda began with finding the optimal ratio to maximize the benzene production, holding the amount of Et_3Al constant and varying the $TiCl_4$ content. In this way, he determined that a benzene yield of ca. 70% could be reached when using Al/Ti ratios of 1–1.5. He then showed that acetylene-$^{14}C_2$ and acetylene-$^2H_2$ could be polymerized at room temperature and low pressure to obtain benzene-$^{14}C_6$ and benzene-$^2H_6$ in yield of about 70%. Lastly, analysis of the isotopic products verified that all of the carbon and hydrogen in the benzene produced originates from acetylene and no exchange processes are involved in the polymerization reaction [35].

This initial study was then followed up with a 1964 investigation of the acetylene polymerization mechanism by following the polymerization of acetylene-$^{14}C_2$ via radio gas chromatography[6] [36]. Acetylene-$^{14}C_2$ was polymerized using a $Et_3Al/TiCl_4$ catalyst mixture (Al/Ti ratio: 2.0) with the residual gas was examined. The gas was primarily ethane, and a small amount of acetylene and a trace amount of ethylene. The level of ^{14}C in the acetylene agreed with the acetylene-$^{14}C_2$ starting

---

[6]Radio gas chromatography is a technique that combines radioisotope tracer techniques with those of gas chromatography.

material, while the ethylene exhibited no detectable ^{14}C and only a minor amount detected in the ethane. Propanol was then added to the polymerization mixture to quench the active catalyst and the gas evolved examined to be primarily ethane with a trace amount of ethylene. The ethylene radioactivity was nearly that of the initial acetylene-$^{14}C_2$ with the ethane again giving a minor, but detectable amount. From these combined results, the various reactions shown in Fig. 5.10 were proposed as potential reactions occurring during polymerization [36].

This was then followed with a more detailed study of the polymerization mechanism in 1966 [37]. In the process, it was found that alkylbenzene was formed as a minor product of acetylene polymerization. When Et$_3$Al was used in combination with TiCl$_4$, ethylbenzene was obtained in addition to simple benzene in a ratio of 0.2–72. If trimethylaluminum (Me$_3$Al) was used, toluene was produced rather than ethylbenzene. It was ultimately concluded that a bond must be formed between the acetylene carbons and the alkyl group of the aluminum co-catalyst before cyclization. This was further confirmed by experiments utilizing ^{14}C and ^2H as tracers, which revealed the both the carbon and hydrogen from the ethyl groups of Et$_3$Al were introduced into benzene. This ultimately led to the proposed reaction steps given in Fig. 5.11 to account for the observed detection of benzene products [37].

Further study of the mechanism of Ziegler-Natta polymerizations in 1967 focused on the stereochemistry of the resulting polyolefin and polyacetylene products [38]. Extending their previous conclusions concerning the formation of alkylbenzene during the polymerization of acetylene, it was proposed that the polyacetylene endgroups could also consist of either alkyl units incorporated from the AlR$_3$ (R = Me or Et) co-catalyst or phenyl groups via a cyclization step that removed the growing polymer from the active catalyst. Efforts to investigate the *cis* versus *trans* relative configuration of the polyacetylene backbone were then undertaken using IR spectroscopy in

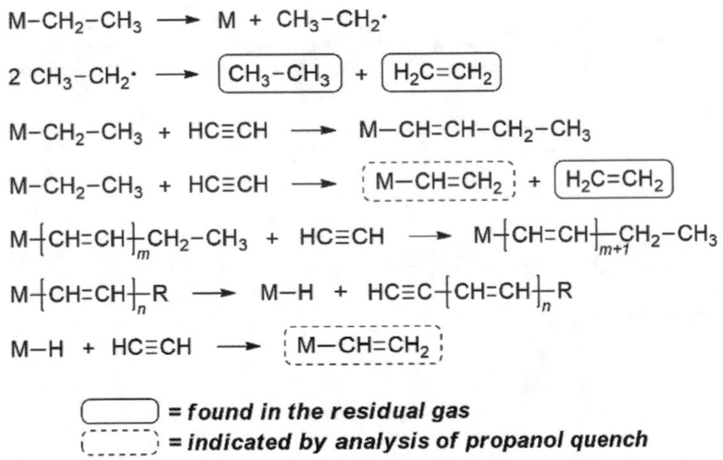

Fig. 5.10 Proposed reaction steps in the Ziegler-Natta polymerization of acetylene

a similar manner to the previous report of Hatano [32]. As with Hatano's previous study, analysis of samples polymerized at temperatures varying from −78 to 80 °C revealed a clear trend of increasing *trans* configuration with increasing temperature. More significantly, however, was that Ikeda also correlated the relative amount of benzene formed at each temperature and found no real temperature effect for benzene formation [38].

It was ultimately concluded that the *cis* configuration was initially formed with each growth step, which could then be thermally converted to the *trans* configuration, although with a relatively large activation energy. The formation of benzene was then thought to be dependent on whether the growing chain takes a *cisoid* structure or a *transoid* structure near the active catalyst [38]. The combination of the results discussed above and the resulting proposed processes thus led to a modified mechanism (Fig. 5.12) in comparison to that previously proposed by Hatano in Fig. 5.9.

[cat]–R + HC≡CH ⟶ [cat]–CH=CH–R

[cat]–CH=CH–R + HC≡CH ⟶ [cat]–CH=CH–CH=CH–R

[cat]–CH=CH–CH=CH–R + HC≡CH ⟶ [cat]–H + C$_6$H$_5$–R

[cat]–H + HC≡CH ⟶ [cat]–CH=CH$_2$

[cat]–CH=CH$_2$ + HC≡CH ⟶ [cat]–CH=CH–CH=CH$_2$

[cat]–CH=CH–CH=CH$_2$ + HC≡CH ⟶ [cat]–H + C$_6$H$_6$

Fig. 5.11 Proposed reaction steps to account for the formation of benzene and alkylbenzene during acetylene polymerization

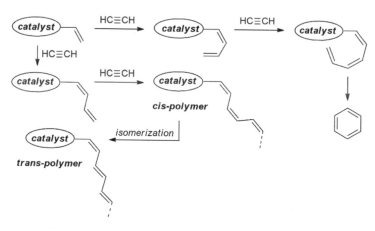

Fig. 5.12 Proposed reaction steps to account for the formation of benzene during acetylene polymerization

In addition to his primary focus of studying the mechanism of olefin and acetylene polymerizations, Ikeda reported the generation of a new catalytic iron complex, $Fe(bpy)_2Et_2$ (where $bpy = 2,2'$-bipyridine) in 1965 [39]. This complex was reported to react with butadiene to give its cyclooligomeric products, but was also capable of catalyzing the production of benzene and polyacetylene from acetylene [39]. The authors stated that a detailed study of this polymerization reaction was to be reported at a later date, but this does not seem to have happened. A more detailed study of the catalyst was reported in 1968, but this second report focused solely on the butadiene reactions [40].

The most well-known work to have come from Ikeda's group, however, was on a new form of polyacetylene reported throughout the 1970s [41–46]. This work was carried out by Ikeda's new assistant, Hideki Shirakawa (b. 1936), who joined Ikeda's group at the Tokyo Institute of Technology in April of 1966 as a research associate [2, 47–49].

5.5 Hideki Shirakawa and Polyacetylene Films

Hideki Shirakawa (Fig. 5.13) was born in Tokyo on August 20, 1936 [2, 17, 48, 49], the third of five children to parents Hatsutarou and Fuyuno [48, 49]. His father was a medical doctor and the family relocated several times after Hideki's birth as his father moved from position to position [48]. In 1944, the family ultimately settled in the small city of Takayama, his mother's hometown located in the middle of Honshu, Japan. It is there that Hideki spent his childhood [2, 48, 49].

He entered Tokyo Institute of Technology in April of 1957 [2, 48], receiving a B.S., M.S. and Ph.D. during his time there from 1957 to 1966 [17]. He focused primarily on applied chemistry during his undergraduate studies [2, 48, 49]. During his final year, he had applied to work in a laboratory working on polymer synthesis,

Fig. 5.13 Hideki Shirakawa (1936) (Reproduced from Ref. [47] with permission of the Royal Society of Chemistry)

but the lab had too many applicants wishing to join. As such, he joined a laboratory working on polymer physics. He was initially reluctant to work on the topic, but in hindsight, he realized the important impact this experience had on his later career [47]. It was during his five years of graduate studies [50] that he was able to return to his primary interest of polymer synthesis, finally receiving a Doctor of Engineering degree in March of 1966 [2, 48, 49, 51] with a dissertation on block copolymers [51]. He then married his wife Chiyoko Shibuya that same year [48, 49]. Following the completion of his graduate studies, he joined the group of Sakuji Ikeda as a research associate at the Tokyo Institute of Technology [2, 17, 47–50, 52]. During that time, he also spent one year over the span of 1966–1967 at the University of Pennsylvania working with Alan G. MacDiarmid and Alan J. Heeger [2, 17, 47–49].

In November of 1979, he moved to the Institute of Materials Science at the University of Tsukuba, where he was appointed associate professor [2, 17, 47–49]. He was later promoted to full professor in October of 1982 and formally retired from the University of Tsukuba at the end of March 2000 [2, 17, 48–50]. Although he still retained the position of professor emeritus, he withdrew from further research or other educational activities [48].

In addition to being awarded the 2000 Nobel Prize in Chemistry with Heeger and MacDiarmid [2, 17, 18, 48, 49], Shirakawa had previously received the 1982 Award of the Japan Society of Polymer Science in May of 1983 [17, 48, 49] and the 1999 Award for Distinguished Service in Advancement of Polymer Science in May of 2000 from the Japan Society of Polymer Science [17, 48]. Finally, he also received the Order of Culture from the Japanese Government in November of 2000 [17, 48].

The critical discovery that polyacetylene could be synthesized as lustrous, silvery films has achieved near mythical status that has been told and retold by many different people over the years, with the story rarely told the same way twice. What is known is that in October of 1967[7] [48, 53], approximately a year and a half after Shirakawa had joined Ikeda's group and only a short time after he had begun research on polyacetylene [48], another researcher under Shirakawa's supervision made an error during the preparation of polyacetylene [47, 48, 52, 54, 55]. This error resulted in the formation of "ragged pieces of a film" [48, 52], rather than the typical black polymer powder usually produced [47, 48, 53–58].

The first point that is rarely agreed upon in the retelling of this event is the identity and nature of the researcher who made the error in question. While the researcher's gender is generally viewed as male [2, 47–49, 54, 55], he has been described by various sources as a "student" [47], "visiting Korean researcher" [47], "visiting scientist" [48], "coworker" [17], "graduate student" [54, 57], "Shirakawa's student" [54], or "foreign student" [55]. It is only in the acknowledgment of his Nobel lecture that Shirakawa finally reveals the name of the researcher to be Dr. Hyung Chick Pyun [16–19, 53].

Hyung Chick Pyun was born on December 23, 1926 in Bongsan county of Hwanghae province, currently located in North Korea [59]. Educated at Kyungdong High

---

[7]Even the year of this discovery has been incorrectly given by several authors, with incorrect dates of "beginning of the 1970s" [47], "the early 1970s" [54], and 1975 [56] all reported.

School, he then entered Seoul National University where he received a B.S. in Chemical Engineering in 1951. With the onset of the Korean War (1950–1953), he began working for the Science Research Institute of the Department of Defense, even before he finished his University studies. He continued there until 1960, when he moved to the Korea Atomic Energy Research Institute (KAERI) [59].[8] The following year, he spent time visiting the University of Kansas [59], where he worked with William E. McEwen (1922–2002) [60]. Upon returning to Korea, he published his first two papers in 1964 [60, 61]. He then received support in 1967 from the International Atomic Energy Agency (IAEA) for a nine-month visit to the Tokyo Institute of Technology [59]. Thus, at the age of 41, he began a period of time working under Sakuji Ikeda in May of 1967. Although it has been said that Pyun had already acquired his doctorate before working in Ikeda's laboratory [52], this is incorrect. After returning to Korea, he finally received a Ph.D. in Nuclear Engineering in 1970 from Seoul National University, based on work he published in 1964 [59, 62]. The only publication that appears connected to his time in Ikeda's lab was a paper published shortly after his return to Korea on the comparison of gamma irradiation versus Ziegler-Natta catalyzed methods for the copolymerization of phenylacetylene and styrene [63]. However, his work turned primarily to polymeric materials and composites after that point [63–69], which seemed to be his research focus for the rest of his career. He retired from KAERI in 1991 [59] and died on March 8, 2018 after an extended illness.

This body of evidence seems to indicate that Pyun was already 17 years into his career at the time of his visit to Ikeda's laboratory and thus not a student. Although he did later receive a Ph.D., his dissertation [64] was not at all related to the work in Ikeda's lab and appears to be connected to his previous visit to the University of Kansas. He would therefore be more correctly considered a visiting researcher or scientist at the time of visit to Japan. Of particular interest is that Pyun was never included as a coauthor on any of the papers by Shirakawa and Ikeda on the generation of the polyacetylene films, although an acknowledgement was given to "Messrs. H. C. Pyun and T. Ito for the preparation of poly(acetylene) films" in the initial 1971 paper [41].

When Pyun made his fortuitous error, he was attempting to polymerize acetylene using methods nearly identical to that previously used by both Natta [15] and Hatano [28] as outlined in Fig. 5.14. In order to understand what had occurred, Shirakawa reviewed the polymerization conditions again and again to ultimately find that the catalyst concentration was nearly 1000 times greater than intended [2, 16–19, 47–49, 52, 54–58]. As a result, the highly concentrated catalyst solution accelerated the rate of the polymerization to the point that, rather than polymerizing in solution to give a black precipitate as was typical, the acetylene polymerized at the air-solvent interface or along the walls of the vessel which had been wetted by the catalyst solution to give the observed silvery films [47, 48, 53].

---

[8]Established in 1959, the name of the Institute changed to the Korea Advanced Energy Research Institute in 1980 before being restored to the original name in 1989. Although Pyun's papers list the city of the Institute as Seoul, it is actually located in Daejeon, ca. 100 miles outside of Seoul.

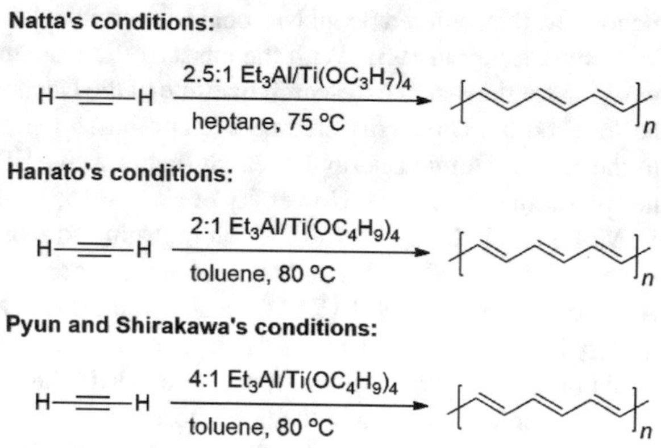

Fig. 5.14 Comparative conditions for the polymerization of acetylene

Natta's conditions:

$$H—\!\!\equiv\!\!—H \xrightarrow[\text{heptane, 75 °C}]{\text{2.5:1 Et}_3\text{Al/Ti(OC}_3\text{H}_7)_4}$$

Hanato's conditions:

$$H—\!\!\equiv\!\!—H \xrightarrow[\text{toluene, 80 °C}]{\text{2:1 Et}_3\text{Al/Ti(OC}_4\text{H}_9)_4}$$

Pyun and Shirakawa's conditions:

$$H—\!\!\equiv\!\!—H \xrightarrow[\text{toluene, 80 °C}]{\text{4:1 Et}_3\text{Al/Ti(OC}_4\text{H}_9)_4}$$

The exact reason for this error is another inconsistent point in the retelling of this event. While most agree that it was the result of miscommunication between Shirakawa and Pyun [2, 47–49, 52, 55, 57, 58], the nature of the miscommunication varies with Shirakawa stating [48]:

> It was an extraordinary unit for a catalyst. I might have missed the "m" for "mmol" in my experimental instructions, or the visitor might have misread it. For whatever reason, he had added the catalyst of some molar quantities in the reaction vessel.

Curiously, his collaborator Alan MacDiarmid gives a quite different account of the event, stating in 2001 [55]:

> I asked him how he [Shirakawa] had made this silvery film of polyacetylene and he replied that this occurred because of a misunderstanding between the Japanese language and that of a foreign student who had just joined his group.

During a 2002 interview with István Hargittai [57], MacDiarmid again gave a very similar account. Interestingly, Shirakawa's other collaborator Alan Heeger gave nearly the same account during an interview with Hargittai in 2004 [58], stating:

> Then he [Shirakawa] had a Korean visitor who misunderstood what he said in Japanese and instead of making the catalyst in the millimolar concentration, he made it in molar concentration and out came something very different.

The chemist and historian István Hargittai, however, confirmed with Shirakawa that Pyun had grown up in Korea during the years that the country was under Japanese occupation (1910–1945), and thus spoke fluent Japanese [52]. This fact, combined with the evidence described above that Pyun was not really a student, casts serious doubt on MacDiarmid's version of the event. While Heeger's account agrees with that of MacDiarmid, Heeger refers to it as a story [58], not fact, thus suggesting that it is not personal knowledge, but something he heard, most likely from MacDiarmid. Still, it is this account of MacDiarmid and Heeger that is typically the most commonly given version in the retelling of this event [52], likely as it gives the more entertaining story.

Still, whatever the reason of the error, refinement of the resulting conditions ultimately allowed Shirakawa to develop methods to reproducibly generate silvery plastic polyacetylene films via polymerization of acetylene on the surface of unstirred, concentrated catalyst solutions [41–45]. Although the initial discovery was made towards the end of 1967, and Shirakawa and Ikeda submitted their first paper on the spectral characterization of the polyacetylene films in November of 1970 [41], they did not publish a detailed synthetic account of the film formation process until 1974 [43]. In a reflective paper published in 1996 [53], Shirakawa addressed the reasons for this six-year delay, stating that the main reason was that the film was not a metallic, but rather a semiconducting polymer. In addition, the film formation was based on interfacial polymerization, which was viewed to be limited and thus not that important a result. Shirakawa stated that this was supported by the fact that when the film formation was reported at domestic Japanese meetings in 1968, little interest was expressed. According to Shirakawa, it was really only after the reports of chemical doping of the film in 1977 with MacDiarmid and Heeger [70], and the corresponding increases in conductivity, that serious interest in these new polyacetylene films developed [53]. However, it should also be pointed out that Shirakawa and Ikeda submitted a patent on the film formation in 1970, which was granted in 1973 [71]. As such, it is also realistic that they were hesitant to report too many details of the process until it was suitably protected by the patent.

Similar to previous efforts by both Hatano [32] and Ikeda [38] on polyacetylene powders, the effect of temperature on the production of the polyacetylene films was then studied in detail. Here, the IR spectra of polyacetylene films and various deuterated derivatives prepared over the range of -100 to $180\,°C$ were analyzed in order to gain insight into the linear structures within the polymer films [41]. Polymer films prepared at temperatures below $-78\,°C$ were described to have a "copper-like luster" [42] (Fig. 5.15) and the IR results led to the conclusion that these films were most consistent with an all-*cis* structure (either *cis-transoid* or *trans-cisoid*, Fig. 5.16). In contrast, films prepared above $150\,°C$ were described as being an "intense black material with a metallic luster" [42] or having a "silvery" [17] appearance on the surface (Fig. 5.15) [72]. As outlined in Table 5.2, it was concluded that both the IR and Raman spectra of these high temperature materials were most consistent with an all-*trans* structure [41–44]. Finally, it was found that the *cis*-form of the polymer could be converted to the *trans*-form via a *cis-trans* isomerization by heating the *cis*-polymer at $200\,°C$ for 10 min [44].

Further study of the polymer film prepared at $-78\,°C$ revealed that it consisted of ca. 98% *cis* content (Table 5.2), while materials prepared at room temperature consisted of ca. 60% *cis* content [16–19, 44]. Isomerization from the *cis* to *trans* forms was found to occur at temperatures as low as $75\,°C$, but quite slowly. Thermograms of the *cis*-polymer revealed an exotherm at $145\,°C$, which was assigned to the irreversible *cis*-to-*trans* isomerization of the polymer backbone [16–19, 44]. Thus, polymerization above $150\,°C$ or heating the *cis*-form above $150\,°C$ both resulted in the isolation of the all-*trans* polyacetylene [44]. Both the *cis* and *trans* forms of the polymer were found to undergo decomposition at temperatures above $325\,°C$ [44].

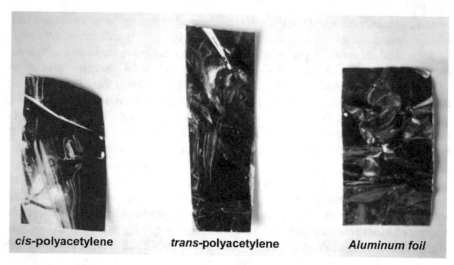

Fig. 5.15 Films of *cis*- and *trans*-polyacetylene in comparison to aluminum foil (Courtesy of Dr. Richard Kaner)

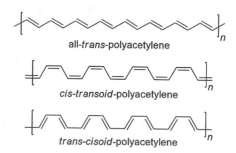

Fig. 5.16   Isomeric structural forms of polyacetylene

The effect of the *cis* versus *trans* content on the electronic properties of poly-acetylene was then reported in 1978 [46]. As shown in Fig. 5.17, the resistivity decreased with increasing *trans* content, although samples with *trans* content greater than 80% did exhibit a tendency to increase in resistivity again as the *trans* content continued to approach 100%. Overall, the all-*cis* samples exhibited conductivities of 10^{-9}–10^{-8} S cm^{-1}, while the all-*trans* samples gave higher conductivities of 10^{-5}–10^{-4} S cm^{-1} [46, 70]. Surprisingly, the values of the all-*trans* films were essentially the same as those previously reported by Watson [27] and Hatano [28] for polyacetylene powders. As Hatano had previously shown the conductivity increased with polymer crystallinity [28], it might be expected that the film would provide increased order and thus a corresponding rise in conductivity. However, this did not seem to be the case and the intrinsic electrical properties of the two forms of polyacetylene appeared to be nearly identical [17]. Finally, characterization of the

Table 5.2 *Cis* versus *trans* content of polyacetylenes prepared at various temperatures [17–19, 43]

Polymerization temperature (°C)	*Cis* content (%)	*Trans* content (%)
−78	98.1	1.9
−18	95.4	4.6
0	78.6	21.4
18	59.3	40.7
50	32.4	67.6
100	7.5	92.5
150	0.0	100.0

Fig. 5.17 Effect of trans
content on polymer
resistivity (Reproduced from
Ref. [46], © 1978 Hüthig &
Wepf Verlag, Basel, used
with permission)

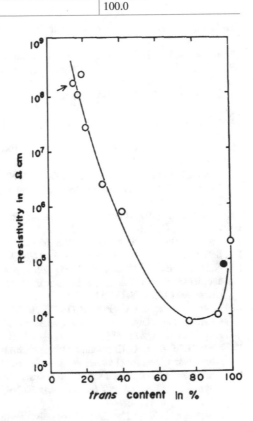

polyacetylene films by X-ray diffraction [43] also gave nearly identical data to that
previously collected by Natta on the powder form [15].

Attempts were then made to convert the polyacetylene films to graphitic materials [16–19]. Differential thermal analysis of *cis*-rich polymer films revealed an
endothermic peak at 420 °C, which was assigned to thermal decomposition, and
thermogravimetric analysis confirmed a corresponding weight loss of 63% at this
temperature [44]. As such, the potential preparation of graphite films via pyrolysis
of the polyacetylene films was not suitable and thus a chlorination/HCl elimination
approach was then investigated. Treatment of polyacetylene with chlorine at room

temperature resulted in the formation of a white film, similar to the material previously reported by Natta [15], which was presumed to be (CHCl)$_n$ [70]. This film was then treated with a basic reagent to eliminate HCl, resulting in the production of a carbon film. However, it was found that little graphitization of the material had occurred even when heated at 2000 °C for several hours [16–19].

References

1. Rasmussen SC (2017) Cuprene: a historical curiosity along the path to polyacetylene. Bull Hist Chem 42:63–78
2. Rasmussen SC (2014) The path to conducting polyacetylene. Bull Hist Chem 39:64–72
3. Morris PJT (1986) Polymer pioneers. Center for the History of Chemistry, Phiadelphia, pp 78–80
4. Ziegler K (1972) In Nobel Lectures, Chemistry 1963–1970. Elsevier Publishing Company, Amsterdam
5. Morawetz H (1985) Polymers. The origins and growth of a science. Wiley, New York, pp 186–192
6. Natta G (1972) In Nobel Lectures, Chemistry 1963–1970. Elsevier Publishing Company, Amsterdam
7. Bawn CEH (1979) Giulio Natta, 1903–1979. Nature 280:707
8. Morris PJT (1986) Polymer pioneers. Center for the History of Chemistry, Phiadelphia, pp 81–83
9. Bamford CH (1981) Giulio Natta—an appreciation. Chem Brit 17:298–300
10. Hargittai I, Comotti A, Hargittai M (2003) Giulio Natta. Chem Eng News 81(6):26–28
11. Porri L (2004) Giulio Natta—his life and scientific achievements. Macromol Symp 213:1–5
12. Schidrowitz P (1950) Giuseppe Bruni 1873–1946. Rubber Chem Technol 23:303–305
13. Natta G, Pino P, Mazzanti G (1955) Polimeri ad elevato peso molecolore degli idrocarburi acetilenici e procedimento per la loro preparazione. Italian Patent 530,753 (15 July 1955); Chem Abst 1958, 52:15128b
14. Natta G, Mazzanti G, Pino P (1957) Hochpolymere von Acetylen-Kohlenwasserstoffen, erhalten mittels Organometall-Komplexen von Zwischenschalenelementen als Katalysatoren. Angew Chem 69:685–686
15. Natta G, Mazzanti G, Corradini P (1958) Polimerizzazione stereospecifica dell'acetilene. Atti Accad Naz Lincei Rend Cl Sci Fis Mat Nat 25:3–12
16. Shirakawa H (2001) The discovery of polyacetylene film: the dawning of an era of conducting polymers. In: Frängsmyr T (ed) Les Prix Nobel. The Nobel Prizes 2000, Nobel Foundation, Stockholm, pp 217–226
17. Shirakawa H (2001) The discovery of polyacetylene film: the dawning of an era of conducting polymers (Nobel lecture). Angew Chem Int Ed 40:2574–2580
18. Shirakawa H (2001) Nobel Lecture: the discovery of polyacetylene film—the dawning of an era of conducting polymers. Rev Mod Phys 73:713–718
19. Shirakawa H (2002) The discovery of polyacetylene film. The dawning of an era of conducting polymers. Synth Met 125:3–10
20. Luttinger LB (1960) A new catalyst system for the polymerization of acetylenic compounds. Chem Ind 1960:1135
21. Luttinger LB (1962) Hydridic reducing agent-group VI11 metal compound. A new catalyst system for the polymerization of acetylenes and related compounds. I. J Org Chem 27:1591–1596
22. Luttinger LB, Colthup EC (1962) Hydridic reducing agent-group VIII metal compound. A new catalyst system for the polymerization of acetylenes and related compounds. II. J Org Chem 27:3752–3756

23. Daniels WE (1962) The polymerization of acetylenes by nickel halide-tertiary phosphine complexes. J Org Chem 29:2936–2938
24. Stille JK, Frey DA (1961) Polymerization of non-conjugated diynes by complex metal catalysts. J Am Chem Soc 83:1696–1701
25. Lenz RW (1990) In memory of John Kenneth Stille. Macromolecules 23:2417–2418
26. Hegedus LS (1990) John K. Stille—biographical sketch. Organometallics 9:3007–3008
27. Watson WH Jr, McMordie WC Jr, Lands LG (1961) Polymerization of alkynes by Ziegler-type catalyst. J Polym Sci 55:137–144
28. Hatano M, Kanbara S, Okamoto S (1961) Paramagnetic and electric properties of polyacetylene. J Polym Sci 51:S26–S29
29. Kanbara S, Hatano M, Hosoe T (1962) 遷移金属アセチルアセトナート-トリエチルアルミニウム系によるアセチレンの重合 (Polymerization of acetylene by transition metal acetylacetonate-triethylaluminum system). J Soc Chem Ind Jpn 65:720–723
30. Ikeda S, Hatano M (1963) Polymerization reaction by organic transition metal complexes. Kōgyō kagaku zasshi 66(8):1032–1037
31. Shimamura K, Hatano M, Kanbara S, Nakada I (1967) Electrical conduction of poly-acetylene under high pressure. J Phys Soc Jpn 23:578–581
32. Hatano M (1967) アセチレン重合体の構造と電気的性質 (Structures and electrical properties of acetylene polymers). Tanso 1967(50):26–31
33. Successive Director, Laboratory for Chemistry and Life Science Institute of Innovative Research, Tokyo Institute of Technology, http://www.res.titech.ac.jp/english/about/director.html. Accessed 1 July 2017
34. Ikeda S (1962) Recent developement in polymer synthesis. J Synth Org Chem Jpn 20:76–86
35. Ikeda S, Akihiro T (1963) Syntheses of benzene-$^{14}C_6$ and benzene-$^2H_6$ using a Ziegler-catalyst. Radioisotopes 12:368–372
36. Ikeda S, Akihiro T, Akira Y (1964) Measurement of C_2 component in acetylene-^{14}C polymerization system by Ziegler catalyst by radio gas chromatography. Radioisotopes 13:415–417
37. Ikeda S, Akihiro T (1966) On the mechanism of the cyclization reaction of acetylene polymerization. J Polym Sci B Polym Lett Ed 4:605–607
38. Ikeda S (1967) チグラー触媒によるエチレンおよびアセチレン重合の立体化学 (Stereochemistry of ethylene and acetylene polymerization by Ziegler catalyst). J Soc Chem Ind Jpn 70:1880–1886
39. Yamamoto A, Morifuji K, Ikeda S, Saito T, Uchida Y, Misono A (1965) Butadiene polymerization catalysts. Diethylbis(dipyridyl)iron and diethyldipyridylnickel. J Am Chem Soc 87:4652–4653
40. Yamamoto A, Morifuji K, Ikeda S, Saito T, Uchida Y, Misono A (1968) Diethyl-bis(dipyridyl)iron. A Butadiene cyclodimerizaton catalyst. J Am Chem Soc 90:1878–1883
41. Shirakawa H, Ikeda S (1971) Infrared spectra of poly(acetylene). Polym J 2:231–244
42. Shirakawa H, Ito T, Ikeda S (1973) Raman scattering and electronic spectra of poly(acetylene). Polym J 4:460–462
43. Ito T, Shirakawa H, Ikeda S (1974) Simultaneous polymerization and formation of polyactylene film on the surface of concentrated soluble Ziegler-type catalyst solution. J Polym Sci Polym Chem Ed 12:11–20
44. Ito T, Shirakawa H, Ikeda S (1975) Thermal cis-trans isomerization and decomposition of polyacetylene. J Polym Sci Polym Chem Ed 13:1943–1950
45. Ito T, Shirakawa H, Ikeda S (1976) ポリアセチレンのシス-トランス組成と固体構造 (Cis-trans composition and solid structure of polyacetylene). Kobunshi Ronbunshu 33:339–345
46. Shirakawa H, Ito T, Ikeda S (1978) Electrical properties of polyacetylene with various cis-trans compositions. Makromol Chem 179:1565–1573
47. Hall N (2003) Twenty-five years of conducting polymers. Chem Commun 1:1–4
48. Shirakawa H (2001) Hideki Shirakawa. In: Frängsmyr T (ed) Les Prix Nobel. The Nobel Prizes 2000, Nobel Foundation, Stockholm, pp 213–216

49. Rasmussen SC (2011) Electrically conducting plastics: revising the history of conjugated organic polymers. In: Strom ET, Rasmussen SC (eds) 100+ years of plastics: Leo Baekeland and beyond. ACS Symposium Series 1080, American Chemical Society, Washington, DC, pp 147–163
50. Shirakawa H (2010) Another role of grants-in-aid: feeding research fruits into society. Kakenhi Essay Series 15:1–5
51. Tokyo Institute of Technology Library's Thesis Database, http://tdl.libra.titech.ac.jp/hkshi/en/recordID/dissertation.bib/TT00000879. Accessed 16 Aug 2013
52. Hargittai I (2011) Risking reputation: conducting polymers. In: Drive and curiosity: what fuels the passion for science. Prometheus Books, Amherst, pp 173–190
53. Shirakawa H (1996) Reflections on "simultaneous polymerization and formation of polyacetylene film on the surface of concentrated soluble Ziegler-type catalyst solution," by Takeo Ito, Hideki Shirakawa, and Sakuji Ikeda, J. Polym. Sci.: Polym. Chem. Ed., 12, 11 (1974). J Polym Sci A Polym Chem 34:2529–2530
54. Kaner RB, MacDiarmid AG (1988) Plastics that conduct electricity. Sci Am 258(2):106–111
55. MacDiarmid AG (2001) Alan G. MacDiarmid. In: Frängsmyr T (ed) Les Prix Nobel. The Nobel Prizes 2000, Nobel Foundation, Stockholm, pp 183–190
56. Miller JS (2000) The 2000 Nobel Prize in Chemistry—a personal accolade. ChemPhysChem 1:229–230
57. Hargittai B, Hargittai I (2005) Alan G. MacDiarmid. In: Candid science V: conversations with famous scientists. Imperial College Press, London, pp 401–409
58. Hargittai B, Hargittai I (2005) Alan J. Heeger. In: Candid science V: conversations with famous scientists. Imperial College Press, London, pp 411–427
59. Byun Hyung Jik, The Chosun Ilbo (chosun.com), http://focus.chosun.com/people/people-01.jsp?id=20494. Accessed 6 Oct 2017
60. Pyun HC (1964) Pre-equilibrium in the Schmidt reaction of benzhydrols. J Korean Chem Soc 8:25–29
61. Pyun HC, Kim JR (1964) Isotopic exchange 5-bromouracil-Br^{82}. J Korean Chem Soc 8:39–42
62. Pyun HC (1970) Pre-equilibrium in the Schmidt reaction of benzhydrols. Ph.D. Dissertation, Seoul National University
63. Pyun HC, Kim J, Lee W-M (1969) Copolymerization of phenyl acetylene with stryrene. J Korean Chem Soc 13:387–393
64. Pyun HC, Kim JR, Lee KH (1972) A study of the preparation of wood-plastic combinations(II). Monomer impregnations and gamma-ray induced polymerizations. J Korean Nucl Soc 4:23–30
65. Kim J, Lee KH, Pyun HC (1972) A study of the preparation of wood-plastic combinations(III). Preparation of wood-plastic combinations by thermal curing method. J Korean Nucl Soc 4:301–305
66. Kim J, Lee KH, Pyun HC (1973) A study of the preparation of wood-plastic combinations(IV). The physical and chemical properties of wood-plastic combinations. J Korean Nucl Soc 5:3–12
67. Pyun HC, Lee KH, Kim J (1973) A study of the preparation of wood-plastic combinations (I). General properties and radiation durabilities of woods. J Korean Nucl Soc 5:150–158
68. Pyun HC, Lee KH, Kim J (1974) Studies on the preparation of mortar-plastic composite. J Korean Nucl Soc 6:73–79
69. Pyun HC, Cho BR, Kwon SK (1975) Preparation of slate-plastic composite. J Korean Nucl Soc 7:9–14
70. Shirakawa H, Louis EJ, MacDiarmid AG, Chiang CK, Heeger AJ (1977) Synthesis of electrically conducting organic polymers: halogen derivatives of polyacetylene, $(CH)_x$. J Chem Soc Chem Commun 578–580
71. Shirakawa H, Ikeda, S (1973) Film and fibers of acetylene high-molecular-weight polymer. JPS4832581 (B1)
72. Saxman AM, Liepins R, Aldissi M (1985) Polyacetylene: its synthesis, doping and structure. Prog Polym Sci 11:57–89

Chapter 6
Doped Polyacetylene

By 1970, various groups were producing polyacetylene as either highly crystalline powders or as metallic looking plastic films, both of which gave maximum conductivities of ca. 10^{-4} S cm^{-1} [1–3]. Still, highly conductive polyacetylene had to wait for the development of polyacetylene materials in either oxidized or reduced forms. Such redox modified conjugated polymers are often referred to as *doped* materials in analogy to the doping of inorganic semiconductors such as silicon [4]. In reality, however, the doping of organic materials has very little in common with the doping of their inorganic counterparts. Although doped polyacetylene is typically viewed to be the first example of such doped, conductive organic polymers, the conductivity of doped polypyrrole was first reported in 1963 [5–7] and the conductivity of doped polyaniline had been reported by 1966 [8–11]. The first example of a doped polyactylene then occurred two years later with a report in 1968 by Donald J. Berets (1926–2002) and Dorian S. Smith (1933–2010) of the American Cyanamid Company [12].

6.1 Berets and Smith

Donald Joseph Berets was born in New York City during July of 1926 and was educated in Cambridge, Massachusetts, where he attended Harvard University. He obtained his A.B. there in 1946, followed by an M.A. in 1947 [13]. His graduate studies focused on physical chemistry and he finished his Ph.D. in 1949 [13–15] with a dissertation titled "Studies on the Detonation of Explosive Gas Mixtures" [15]. Berets then moved to neighboring Massachusetts Institute of Technology (MIT), where he spent a year as a postdoctoral fellow [13]. Afterwards, he moved to Stamford, Connecticut to begin a career in research and development at American Cyanamid Company [13, 14]. His work at American Cyanamid focused on heterogeneous catalysts for oil refining, auto emissions control, and synthetic fuels.

© The Author(s) 2018
S. C. Rasmussen, *Acetylene and Its Polymers*, SpringerBriefs in Molecular Science,
https://doi.org/10.1007/978-3-319-95489-9_6

Berets retired from American Cyanamid in 1986, after which he formed the Chemists Group, an affiliation of approximately 150 independent consultants in chemistry and chemical engineering [13, 14]. Throughout his career, Berets was very active in the American Chemical Society (ACS), where he served as the councilor from the ACS Western Connecticut Section for 35 years, chaired the ACS Subcommittee on Professional Standards & Ethics from 1986 to 1989, chaired ACS's Council Committee on Professional Relations in 1990, and authored the "Chemist's Code of Conduct" adopted by ACS in 1994 [13]. In January 2002, he received the ACS Presidential Plaque and the Western Connecticut Section's Julius Kuck Service Award for his long service to ACS. He died in Stamford on February 2, 2002 at the age of 75 [13].

Beret's colleague and coworker, Dorian Sevcik Smith, was born in 1933 and grew up in Winthrop Harbor, Michigan [16]. He then moved to Normal, Illinois, to attend Illinois State Normal University (now Illinois State University) [16, 17], where in addition to his coursework, he played football and was part of the University's sole undefeated football team in 1950 [16]. As a result, the 1950 Redbirds team was inducted into the Illinois State Athletics Percy Hall of Fame in 1990 [18]. After completing a B.S. in Education in 1953 [17], the Teachers College Board of the State of Illinois appointed Smith as a faculty assistant for the 1953–1954 academic year [19]. He then continued his education with graduate studies in chemistry at the University of Illinois at Urbana-Champaign [16, 17]. There, he studied under Therald Moeller (1913–1997), receiving a M.S. in 1956 [17]. He completed his Ph.D. two years later in 1958, with a dissertation titled "Observations on the Rare-Earths: Chemical and Electrochemical Studies in Non-Aqueous Solvents" [17].

Smith spent the next 10 years as an industrial chemist [16], first at the American Cyanamid Company, and then briefly at the Enjay Chemical Company in New York [12], before becoming a financial analyst for various firms, including Donaldson, Lufkin and Jenrette, Chemical Bank and Yamaichi International, where he was Director of Research [16]. After spending the majority of his working life in New York City, he retired in 1996 to Wilmington, North Carolina. At the age of 77, Smith passed away peacefully at his home on December 4, 2010 [16].

Approximately a month before Shirakawa and Pyun accidently produced the first polyacetylene films,[1] Berets and Smith submitted a paper detailing the effects of additives on the conductivity of polyacetylene powders [12]. These efforts began with an investigation of the effect of oxygen impurities on the conductivity of polyacetylene pressed pellets, which confirmed samples with lower oxygen content gave lower resistivity. The sample with the least oxygen content (0.7%) gave a resistivity of 7.5×10^5 Ω cm, in good agreement with that previously reported by Hatano for polyacetylene of medium crystallinity [2]. In the process, however, they observed an interesting phenomenon associated with the presence of oxygen in contact with the polyacetylene sample [12]:

> On admission of 150 mm pressure of oxygen to the measuring apparatus (normally evacuated or under a few cm pressure of He gas), the resistivity of polyacetylene decreased by a factor

---

[1] See Chap. 5.

of 10. If the oxygen was pumped off within a few minutes and evacuation continued at 10^{-4} mm pressure for several hours, the original electrical properties of the specimen were restored.

They went on to conclude that oxygen is first adsorbed in a reversible manner that results in a reduction in the resistivity. Oxygen, however, ultimately reacts with the polymer irreversibly resulting in the typically observed increase in resistivity. This initial reversible process was the first recognition of what is now commonly referred to as the oxygen doping of conjugated polymers and it is only the enhanced reactivity of polyacetylene in comparison to other conjugated polymers that causes the second irreversible process. The predominance of this second process had masked the initial process in other studies and thus had not been previously observed.

Following this, further gases were then investigated to determine any effects on the polymer conductivity. These gases included BF_3, BCl_3, HCl, Cl_2, SO_2, NO_2, HCN, O_2, ethylene, NH_3, CH_3NH_2, H_2S, acetone, $(C_2H_5)_2CO$, N_2, and He [12]. Many of these gases had essentially no effect. It was found, however, that electron acceptors (BF_3, BCl_3, Cl_2, SO_2, NO_2, and O_2) and the two acids (HCl and HCN) all resulted in a decrease in resistivity (i.e. an increase in conductivity). As previously observed for O_2, strongly oxidizing gases (O_2, Cl_2, and NO_2) ultimately resulted in chemical reaction with the polymer. In contrast, electron donors (NH_3 and CH_3NH_2) had the opposite effect on resistivity. The greatest increase in conductivity was observed using BF_3, causing an increase of three orders of magnitude (to ~ 0.0013 S cm^{-1}). Berets and Smith provided the following explanation for these results [12]:

> The effect on conductivity of the adsorbed electron-donating and electron-accepting gases is consistent with the p-type nature of the specimens... If holes are the dominant carriers, electron donation would be expected to compensate them and reduce conductivity; electron acceptors would be expected to increase the concentration of holes and increase conductivity; this is observed.

Although Berets and Smith didn't completely understand the effect of the gaseous additions, they quite clearly state that the "electrical conductivity...depended on the extent of oxidation of the samples" [12]. These results, however, did not seem to generate much interest and Berets and Smith never followed up this work with any additional studies. In fact, this work was published just shortly before Smith left chemistry to become a financial analyst and it seems to be his only paper published after completing his Ph.D.

Although Hideki Shirakawa was well aware of the report of Berets and Smith [3, 20, 21], their work has never been referenced by Shirakawa, Alan MacDiarmid, or Alan Heeger as a factor contributing to the later, more successful doping of polyacetylene in 1977 [22, 23]. Rather, it is typically the related work of MacDiarmid and Heeger on poly(sulfur nitride) [24–27] that is cited as the motivation behind efforts to dope polyacetylene with halogens [22, 28–30].

Fig. 6.1 Alan G. MacDiarmid (1927–2007) (left) and Alan J. Heeger (b. 1936) (right). (Reproduced from Ref. [31] with permission of the Royal Society of Chemistry)

6.2 MacDiarmid, Heeger, and $(SN)_x$

Alan Graham MacDiarmid (Fig. 6.1) was born in Masterton, New Zealand on April 14, 1927 [28–31]. Coming from a very economically-challenged family, his family had soon moved to a suburb of Wellington, where it was believed that jobs were more plentiful in the 1930s [28]. After his father had retired, MacDiarmid was forced to leave high school at age 16 in order take a part-time job as janitor and lab boy in the chemistry department of Victoria University College [28, 29] (now Victoria University of Wellington), a constituent college of the University of New Zealand at the time [29]. The position consisted primarily of washing dirty labware, sweeping floors, and preparing demonstration materials for the faculty, which allowed him time to apply himself as a part-time student [28, 29]. As such, he was able to complete his B.Sc. in chemistry in 1948 [29], after which he was promoted to the position of demonstrator [28]. About this same time, he began studying the chemistry of S_4N_4 for his M.Sc. thesis under Mr. A. D. Monro, the lecturer in first-year chemistry [28, 29]. This led to the publication of his first paper in 1949 [32] and completed his M.Sc. in chemistry the following year [29, 33].

MacDiarmid then received a Fulbright Scholarship in 1950 to attend the University of Wisconsin [28, 29, 34]. There, he studied inorganic chemistry under Norris F. Hall [28, 29], earning a M.S. in 1952 [29] and a Ph.D. in 1953 [29, 31, 33, 35] with a dissertation entitled "Isotopic Exchange in Complex Cyanide—Simple Cyanide Systems" [35]. While still at Wisconsin, he obtained a Shell Oil graduate scholarship, which allowed him to attend Cambridge University in England [29]. There, he studied silicon hydrides under Harry J. Emeléus (1903–1993) [29, 33] and completed his second Ph.D. in 1955 [29, 31, 33, 36] with a dissertation entitled "The chemistry of some new derivatives of the silyl radical" [36]. He then held a brief appointment as Assistant Lecturer at the University of St. Andrews in Scotland while also joining the Department of Chemistry at the University of Pennsylvania that same year in 1955 [29, 33].

MacDiarmid maintained his position at the University of Pennsylvania until the end of his career, but he also held positions at the University of Texas at Dallas and Jilin University in China. At Dallas, he became the James Von Ehr Chair of Science and Technology and professor of chemistry and physics in 2002, and he became professor of chemistry at Jilin University in 2004 [29]. In addition to being awarded the 2000 Nobel Prize in Chemistry with Shirakawa and Heeger [29, 34], MacDiarmid also received the Chemistry of Materials Award from the American Chemical Society in 1999 and the Rutherford Medal from The Royal Society of New Zealand in 2001 [29]. MacDiarmid continued to work until his death on February 7, 2007 at the age of 79 [34].

MacDiarmid's colleague in physics, Alan Jay Heeger (Fig. 6.1), was born in Sioux City, Iowa on January 22, 1936 [37, 38]. He spent his early years in nearby Akron, and it was there that he began his primary education [39]. His father died when he was only nine [37, 38], and thus in 1947 [39, 40] his family moved to Omaha, Nebraska, in order to be closer to his mother's family [37, 38, 40]. There, he attended Omaha Central High School [41] and later attended the University of Nebraska [37, 38]. His initial goal was to become an engineer, but he abandoned engineering after the first semester to pursue dual studies in physics and mathematics [37, 38, 42]. After completing his B.S. in 1957, he was accepted into the graduate programs at Cornell University and the University of California at Berkeley, and he decided to attend Cornell. He only remained at Cornell for one year, however, as he did not have the funds needed to continue his education there [43]. He then heard that the Lockheed Missiles and Space Division in California offered programs which allowed one to work part time while also continuing graduate studies, with the tuition costs covered by Lockheed. Thus, he applied and was accepted into the Lockheed program, although he still needed to be accepted into a nearby graduate program. As such, he applied to the physics programs at both Stanford University and Berkeley, but was only accepted into the Berkeley program [43].

Heeger then began graduate studies at Berkeley in 1958, while also working part time for Lockheed Missiles and Space Division in Palo Alto, California, where he carried out work on silicon solar cells [37, 38, 43]. Thus, three days a week he would drive to Berkeley to attend morning classes, before returning to Palo Alto for work in the afternoon. This schedule quickly took its toll, however, and he eventually moved to Berkeley to pursue his studies on a full-time basis [37, 43]. For his graduate research, his initial goal was to do theoretical work under Charles Kittel (b. 1916). Although Kittel was willing to take him on as a student, Kittel recommended instead that he consider working with someone who does experimental work in close interaction with theory. Heeger decided to follow his advice and thus he joined the research group of Alan Portis (1926–2010) [37, 38, 43]. He completed his Ph.D. in 1961 with a dissertation entitled "Studies on the Magnetic Properties of Canted Antiferromagnets" [44].

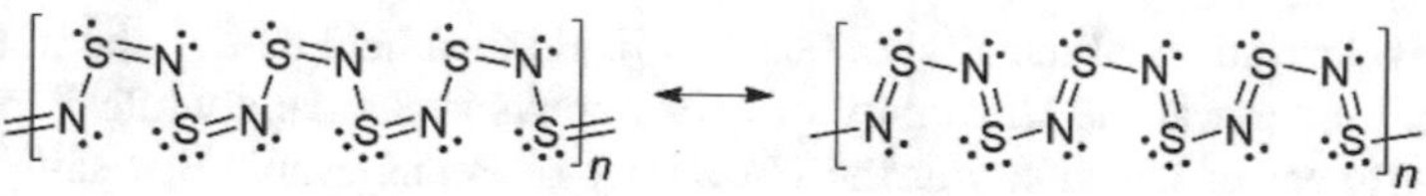

$$\left[\substack{\overset{\cdot\cdot}{S}=N\cdot \\ \Vert \\ \mathbin{\mskip-3mu}N\cdot} \quad \substack{\overset{\cdot\cdot}{S}=N\cdot \\ \Vert \\ \cdot\overset{\cdot\cdot}{S}=N\cdot} \quad \substack{\overset{\cdot\cdot}{S}=N\cdot \\ \Vert \\ \cdot\overset{\cdot\cdot}{S}=N\cdot} \quad \substack{ \\ \\ \cdot\overset{\cdot\cdot}{S}}\right]_n \longleftrightarrow \left[\substack{\overset{\cdot\cdot}{S}-\overset{\cdot\cdot}{N} \\ \Vert \\ N\cdot} \quad \substack{\overset{\cdot\cdot}{S}-\overset{\cdot\cdot}{N} \\ \Vert \\ \cdot\overset{\cdot\cdot}{S}-\overset{\cdot\cdot}{N}} \quad \substack{\overset{\cdot\cdot}{S}-\overset{\cdot\cdot}{N} \\ \Vert \\ \cdot\overset{\cdot\cdot}{S}-\overset{\cdot\cdot}{N}} \quad \substack{ \\ \\ \cdot\overset{\cdot\cdot}{S}}\right]_n$$

Fig. 6.2 Poly(sulfur nitride)

After completing his Ph.D., Heeger stayed on at Berkeley for six months as a postdoctoral research fellow [43], before joining the Physics Department at the University of Pennsylvania in January of 1962 [37], working initially on the metal physics of tetrathiafulvalene-tetracyanoquinodimethane (TTF-TCNQ) [37, 38]. In 1982, Heeger moved to the University of California, Santa Barbara (UCSB), where he still actively pursues research in conjugated materials. In addition to his faculty position as Professor of Physics, Heeger also served as the founding Director of UCSB's Institute for Polymers and Organic Solids [45].

The collaboration between the two Pennsylvania (Penn) colleagues began in 1975, after Heeger had become intrigued by reports of poly(sulfur nitride), $(SN)_x$, also known as polythiazyl (Fig. 6.2) [28–30, 37, 38, 46]. First reported in 1973 by Mortimer M. Labes at nearby Temple University [47–49], Heeger was eager to study the material due to its reported metallic properties, which was highly unusual for a one-dimensional polymer containing no metals. Learning that MacDiarmid had some experience with sulfur nitride chemistry, Heeger approached him about working together on a study of this new material [28–30, 37, 38]. Their formal collaboration then began with development of the first reproducible preparation of analytically pure $(SN)_x$ via the solid-state polymerization of S_2N_2, resulting in the formation of a lustrous golden material [24, 50]. This was then continued in 1976 with a short report on the stability of $(SN)_x$ to O_2 and H_2O [51], followed by a second paper on the crystal structures of both S_2N_2 and $(SN)_x$ [52], as well as the structural changes occurring during polymerization, and then finally a third paper detailing the polymer's electronic properties [26].

The polymeric material was found to give room temperature conductivities of 1.2–3.7×10^3 S cm^{-1} [26], in comparison to values previously reported by Labes and coworkers that ranged from 1.0×10^2 to 1.73×10^3 S cm^{-1} [47, 48]. In addition, the temperature dependence of the conductivity was found to be consistent with that for ordinary metals [26]. Finally, following up on previous reports that $(SN)_x$ reacted with halides, the polymer was treated with bromine vapor to produce $(SNBr_y)_x$ derivatives as shiny black crystals with a lustrous blue-purple tinge [27]. The ratio of bromine content (i.e. y) was determined to range between 0.38 and 0.42, and the bromide derivatives gave an average room temperature conductivity of 3.8×10^4 S cm^{-1}, a 10-fold increase in comparison to the original $(SN)_x$. As a result of its metallic conductivity and the lack of metallic elements in its composition, poly(sulfur nitride) joined intercalated graphites in the growing class of materials known as *synthetic metals* [53, 54].

6.3 Development of Doped Polyacetylene Films

In 1975 [28, 30], shortly after the beginning of the collaboration with Heeger [37, 38, 46], MacDiarmid traveled to Japan to spend three quarters of a year as a visiting professor at Kyoto University [28, 29]. Towards the end of his visit in 1976 [55], he was asked to give a lecture at the Tokyo Institute of Technology, were in addition to his silicon work, he also described the ongoing work on $(SN)_x$ [28, 31, 46, 56]. After the lecture he was invited to have tea with the head of the chemistry department and it there that he first met Hideki Shirakawa [28, 29, 46]. As Shirakawa had not attended MacDiarmid's lecture [29, 46], MacDiarmid showed him his sample of golden $(SN)_x$ over tea [28–30, 56]. After seeing MacDiarmid's golden material, Shirakawa said that he had a similar material and returned to his lab to retrieve a sample of his silver-colored polyacetylene film to show MacDiarmid [29]. The film captured MacDiarmid's interest and after returning to Penn, he inquired into the possibility of supplemental funding in order to bring Shirakawa to the US to work with them on polyacetylene. ONR Program Officer Kenneth Wynne agreed to support Shirakawa's visit [29–31, 37, 38, 46] and thus he began working with MacDiarmid and Heeger as a visiting scientist in September of 1976 [55, 57, 58].

Upon arriving at Penn, Shirakawa and MacDiarmid began their polyacetylene studies by working to improve the purity of the films in order to increase their conductivity [28]. As discussed above, previous studies by Smith and Berets [12] had shown that decreasing oxygen content did increase conductivity and thus limiting other impurities could possibly further increase the film's conductivity. Eventually, films were successfully produced with purities as high as ca. 98.6% [29]. Surprisingly, however, it was found that the film's conductivity decreased with enhanced purity [28, 29]. In the process, the temperature-dependence of the conductivity was also investigated to determine whether polyacetylene had the temperature profile of a metal or semiconductor [29]. As both William H. Watson [1] and Masahiro Hatano [2] had previously reported in 1961 that the conductivity of polyacetylene powders exhibited a temperature dependence typical of semiconductors,[2] it is unclear if it was thought that the film might have a different response in comparison to the pressed pellets, or if they were just unaware of the previous work at the time. However, as Shirakawa had previously referenced both of these works in his earlier papers [21, 59], he at least must have been aware of the known temperature dependence of polyacetylene powers.

Based on the observed relationship between increasing purity and decreasing conductivity, it was proposed that the impurities in the polyacetylene films were perhaps acting as dopants which thus increased the measured conductivity [28, 29]. As such, it was reasoned that the purposeful addition of bromine may result in enhanced conductivity in a manner similar to that previously observed by Heeger and MacDiarmid in the case of $(SN)_x$ [27, 28]. This reasoning was also supported by in situ infrared (IR) measurements previously carried out by Shirakawa and Ikeda during the treatment of polyacetylene films with chlorine. These measurements revealed a

---

[2]See Chap. 5.

dramatic decrease in IR transmission [58, 60–62], which suggested that the material might have unusual electronic properties [30].

Shirakawa and Dr. Chwan K. Chiang, a postdoctoral fellow of Heeger [57], finally performed the landmark experiment on November 23, 1976 [57, 58, 60–62]. Using a four-point probe, the room temperature conductivity of a *trans*-polyacetylene film was measured while being exposed to bromine vapor [22, 57]. Upon the addition of 1 Torr of bromine, the conductivity rapidly increased (from 10^{-5} to 0.5 S cm^{-1}), resulting in a change of approximately four orders of magnitude within only 10 min [22]. The measurements were then repeated using iodine in place of bromine to give an increase in conductivity up to 38 S cm^{-1}. The highly conducting materials resulting from the treatment of halides were proposed to be charge-transfer π-complexes similar to those believed to be formed during the halogenation of olefins. This treatment of polyacetylene with halides was collectively referred to as "chemical doping" [22].

Following the initial May 1977 communication [22], further results were reported in October of the same year [23]. By optimizing the rate of the iodine treatment, the conductivity of *trans*-polyacetylene was increased up to 160 S cm^{-1}. Better yet, it was found that the treatment of the polymer films with AsF$_5$ gave conductivities of 220 S cm^{-1} for the *trans* isomer, with even higher values (560 S cm^{-1}) for the *cis* isomer. They went on to show that ca. 1% of the AsF$_5$ dopant was needed to achieve metallic conductivity and above this dopant concentration, the doped polyacetylene exhibited a temperature dependence consistent with a disordered one-dimensional metal [23].

In March of 1978, two additional reports were published [63, 64]. The first of these demonstrated that polyacetylene could be doped with electron-donating species such as sodium and that such sodium-doped films could be compensated via oxidative doping with I$_2$ [63]. The second report summarized the group's results to date on halogen-doped polyacetylene, as well as providing additional detail on the temperature-dependent conductivities of these materials [64]. As part of this, the doping of polyacetylene with iodine was repeated using *cis*-polyacetylene resulting in conductivity values above 500 S cm^{-1}, in comparison to the previous value of 160 S cm^{-1} for the *trans* isomer. A third publication in April then reported the characterization of both the *cis*- and *trans*-isomers by ^{13}C NMR [65].

In the fall of 1978, a more expansive report of the various doped polyacetylenes and their corresponding maximum conductivities was then published (Table 6.1) [66]. While doping with sodium had been previously reported [63], the previous work had only included ratios of conductivity and this new study thus included the first reported conductivity value of 80 S cm^{-1} for sodium-doped polyacetylene, as well as the results for films doped with HBr and various mixed halogen species (ICl, IBr). The most conductive forms continued to be that of the AsF$_5$-doped films with a maximum conductivity of 560 S cm^{-1}. Adjusting the conductivity relative to the material's density, the per weight conductivity of these AsF$_5$-doped films was determined to be roughly one half that of copper metal [66].

Shirakawa returned to the Tokyo Institute of Technology in 1978, but still published two final papers as collaborations between himself and Ikeda in Japan with Heeger and MacDiarmid at Penn [67, 68]. The first of these papers was published

Table 6.1 Conductivity of doped polycrystalline polyacetylene films [66]

Material[a]	σ (S cm^{-1})[b]	Material[a]	σ (S cm^{-1})[b]
trans-[(CH)(HBr)$_{0.04}$]$_x$	7×10^{-4}	trans-[CHI$_{0.20}$]$_x$	160
trans-[CHCl$_{0.02}$]$_x$	1×10^{-4}	cis-[CH(IBr)$_{0.15}$]$_x$	400
trans-[CHBr$_{0.05}$]$_x$	5×10^{-1}	trans-[CH(IBr)$_{0.12}$]$_x$	120
trans-[CHBr$_{0.23}$]$_x$	4×10^{-1}	trans-[CH(AsF$_5$)$_{0.03}$]$_x$	70
cis-[CH(ICl)$_{0.14}$]$_x$	50	trans-[CH(AsF$_5$)$_{0.10}$]$_x$	400
cis-[CHI$_{0.25}$]$_x$	360	cis-[CH(AsF$_5$)$_{0.14}$]$_x$	560
trans-[CHI$_{0.22}$]$_x$	30	trans-[Na$_{0.28}$(CH)]$_x$	80

[a]Cis and trans prefixes denote the initial pre-doped form of the polyacetylene sample. [b]At 25 °C

in the summer of 1978 and reported optical absorption and reflection measurements carried out on stretch-oriented films of polyacetylene and its doped products. It was concluded that the data were consistent with a model of (CH)$_x$ as a direct bandgap semiconductor with a bandgap of 1.4–1.6 eV. Doping of the films resulted in a reduced intensity of the interband transition and a new absorption within the gap. The anisotropic reflectance from partially aligned films and the increased optical anisotropy upon doping was concluded to provide evidence for microscopic anisotropy in both the semiconducting and metallic state of the material [67].

The study of the anisotropic nature of stretch-oriented films was then continued in a second paper published in April of 1979, which focused on the electrical conductivity of the oriented material [68]. This study compared the electrical properties of the as-grown AsF$_5$-doped film, with analogous stretch-oriented samples such that the stretched axis had been elongated by a factor of either two or three. In this way, it was found that the conductivity parallel to the elongation increased from a value of 300 S cm^{-1} for the as-grown film, to ca. 1500 S cm^{-1} for the intermediately elongated films, to an excess of 2000 S cm^{-1} for the most elongated films. The conductivity perpendicular showed a minor decrease with stretch elongation. The maximum conductivity parallel to the elongated axis of the stretch-orientation was found to give an average value of 2150 S cm^{-1} [68].

6.4 Continued Study of Doped Polyacetylene Films

After Shirakawa returned to Japan, MacDiarmid and Heeger continued to report on additional studies of polyacetylene films without him. The first of these papers was published in November of 1978, which focused on computational modeling of the electronic structure of polyacetylene in order to interpret its optical specta [69]. This was then followed by a number of papers in 1979, beginning with a report in February on the ESR characterization of trans-polyacetylene films in its native and doped forms [70]. The concentration of unpaired electrons in the undoped material was found to be 1.44×10^{19} e$^-$/g, which is consistent with the low crystalline powder samples report by Hatano (1.1×10^{19} e$^-$/g) [2], although MacDiarmid and Heeger never made this comparison. Changes in the ESR spectra upon AsF$_5$ doping were then concluded to be indicative of the generation of free carriers, which would be consistent with the observed rise in conductivity. This was then followed with a second February 1979 paper on the electrochemical doping of polyacetylene films in various electrolyte solutions [71]. Of these, the use of [Bu$_4$N]ClO$_4$ resulted in the highest conductivity to date, with values up to 970 S cm^{-1}.[3]

After publishing a review of the work on polyacetylene films in April of 1979 [72], the duo then continued their studies with a May report on the photoelectron spectra of AsF$_5$-doped polyacetylene [73]. From the results, it was concluded that at least a portion of the arsenic fluoride lie near the surfaces of the (CH)$_x$ fibrils, but the results were inconclusive in determining the composition of the arsenic fluoride species in the doped film.

This was then followed with a June report detailing new methods of producing low density forms of polyacetylene via a gel intermediate [74]. It was concluded that the fibril morphology of the previous films was preserved, although with larger fibril sizes. Overall, the lower density samples produced lower conductivities with similar doping levels in comparison to the higher density films. Additional papers detailed the characterization of the magnetic susceptibility [75], optical and IR properties [76], and morphological studies [77] of neutral and doped polyacetylene were also reported in the following months of 1979.

6.5   Naarmann and Efforts to Maximize Conductivity

Although Shirakawa, MacDiarmid, and Heeger reported conductivities greater than 2000 S cm^{-1} in stretch-oriented polyacetylene [68], theory had predicted the possibility of conductivities greater than 20,000 S cm^{-1} [75]. Efforts to achieve higher conductivities were then reported by Herbert Naarmann (Fig. 6.3) and academic coworkers beginning in 1986 [78]. Naarmann was initially educated at the West-fälische Wilhelms-Universität Münster (commonly referred to as the University of Münster), where he studied organic and physical chemistry from 1951 to 1954 [79]. He then continued his education at Julius-Maximilians-Universität Würzburg (commonly referred to as the University of Würzburg), were he recieved his Ph.D. in 1959 for a thesis entitled "On the identification of constituents of the poison of some bird spiders" [80]. Naarmann then joined BASF in June of 1960, where his work focused on the structure-function relationships in organic and polymeric materials.

Naarmann's initia efforts involved the polymerization of acetylene over an aged Ti(OBu)$_4$-A1Et$_3$ catalyst mixture in silicone oil as a more viscous medium. This resulted in the production of a glossy golden polyacetylene film on the surface of the

---

[3]This was quickly exceeded by the conductivity of the stretch-oriented, AsF$_5$-doped films reported in April of that same year [68].

Fig. 6.3 Herbert Naarmann.
(Reproduced from Ref. [79]
with permission of John
Wiley and Sons)

catalyst medium, which was washed repeatedly with toluene, 6% HCl in methanol, and methanol. This purified polymer was then doped with a $FeCl_3$-nitromethane solution to achieve conductivities as high as 2340 S cm^{-1} [78].

This was then followed by a paper in January of the following year [81] that attempted to probe the effect of polymerization conditions on the properties of the resulting films. In particular, these efforts focused on the morphology of the resulting films, concluding that the density of the film produced was directly related to the catalyst concentration, with higher concentrations giving more dense films [81]. As a result, very low concentrations result in powder products due to the availability of the acetylene gas to penetrate the catalyst solution, while high concentrations cause the rapid formation of a thick film that limits diffusion of acetylene to the surface. At select intermediate concentrations, a gel is formed rather than a true film. Examination of the products via scanning electron microscope revealed that all materials consisted of a fibrillar structure which was independent of the macroscopic morphology.

Two additional papers then investigated the effects of different catalyst mixtures, with particular emphasis on the investigation of $Ti(OBu)_4$-butyl lithium (BuLi) mixtures [82, 83]. While these did produce high quality and highly conductive polyacetylene, the previous $Ti(OBu)_4$-AlEt$_3$ still provided the best results. As shown in their earlier study [78], the use of silicone oil as a liquid media always gave higher quality films then the traditional use of toluene [83].

This was followed by an additional report in the spring of 1987 that detailed the further modification of these approaches to the production of high quality polyacetylene films [84]. Here, the silicone oil $Ti(OBu)_4$-AlEt$_3$ medium was applied to a flat carrier consisting of a stretchable polymer-supporting material (i.e. polyethylene or polypropylene substrates), after which acetylene was polymerized over the surface to generate a black, homogeneous film. This film could either be washed and removed, or it could be stretched prior to the removal from the supporting substrate.

The unstretched polyacetylene gave conductivities of 2000 S cm^{-1} upon doping with I$_2$, which could then be increased up to 18,000 S cm^{-1} upon stretch orientation.

Reported in the same paper [84], it was also found that even higher conductivities could be achieved through a hybrid catalyst mixture that combined the two primary catalyst systems previously investigated. In this approach, small amounts of a BuLi solution were added to the Ti(OBu)$_4$-AlEt$_3$ catalyst medium. The action of the BuLi was believed to be as a reducing agent, which enhanced the defect-free nature of the resulting polyacetylene films. Unstretched I$_2$-doped films produced in this way provided conductivities of 5000 S cm^{-1}, which could then be increased as high as 170,000 S cm^{-1} upon stretch orientation. Such high quality doped polyacetylene films were now exhibiting conductivities per unit weight that was greater than that of metals such as copper (i.e. ca. 150,000 S cm^{-1} g^{-1} vs. ca. 72,000 S cm^{-1} g^{-1} for Cu). This new form of polyacetylene then became known as *Naarmann-type polyacetylene* (or N-(CH)$_x$) [84].

A collaborative study with Heeger was then reported in June of 1987, which probed the temperature- and pressure-dependance of the conductivity of stretch-oriented polyacetylene generated via the initial Ti(OBu)$_4$-AlEt$_3$ catalyst [85]. The I$_2$-doped material gave a room temperature parallel conductivity above 20,000 S cm^{-1}, which did diminish with temperature, but was still above 9000 S cm^{-1} at 0.5 K. As such, it was concluded that the magnitude and temperature independence of the parallel conductivity below 1 K implied genuine metallic behavior and ruled out transport via hopping among strongly localized states. Still, the increase in conductivity with temperature implied that phonon-assisted transport was involved.

Naarmann and his coworkers continued to study the electronic properties of poly-acetylene for the next several years [86–91], but no further increases in the conductivity were achieved. Nevertheless, the impressive conductivity levels achieved by Naarmann in the late 1980s still remain the highest conductivity reported for any conducting organic polymer. Thus, while focus in the field of conjugated materials has moved away from polyacetylene, polyacetylene will always remain a landmark conductive material for the field.

References

1. Watson WH Jr, McMordie WC Jr, Lands LG (1961) Polymerization of alkynes by Ziegler-type catalyst. J Polym Sci 55:137–144
2. Hatano M, Kanbara S, Okamoto S (1961) Paramagnetic and electric properties of polyacetylene. J Polym Sci 51:S26–S29
3. Shirakawa H, Ito T, Ikeda S (1978) Electrical properties of polyacetylene with various *cis-trans* compositions. Makromol Chem 179:1565–1573
4. Rasmussen SC (2017) Early history of conductive organic polymers. In: Zhang Z, Rouabhia M, Moulton SE (eds) Conductive polymers: electrical interactions in cell biology and medicine. CRC Press, Boca Raton, Chapter 1
5. McNeill R, Siudak R, Wardlaw JH, Weiss DE (1963) Electronic conduction in polymers. Aust J Chem 16:1056–1075

6. Bolto BA, Weiss DE (1963) Electronic conduction in polymers. II. The electrochemical reduction of polypyrrole at controlled potential. Aust J Chem 16:1076–1089
7. Bolto BA, McNeill R, Weiss DE (1963) Electronic conduction in polymers. III. Electronic Properties of polypyrrole. Aust J Chem 16:1090–1103
8. Jozefowicz M, Yu LT (1966) Relations entre propriétés chimiques et électrochimiques de semi-conducteurs macromoléculaires. Rev Gen Electr 75:1008–1013
9. Yu LT, Jozefowicz M (1966) Conductivité et constitution chimique pe semi-conducteurs macro-moléculaires. Rev Gen Electr 75:1014–1018
10. De Surville R, Jozefowicz M, Yu LT, Perichon J, Buvet R (1968) Electrochemical chains using protolytic organic semiconductors. Electrochim Acta 13:1451–1458
11. Jozefowicz M, Yu LT, Perichon J, Buvet R (1969) Proprietes nouvelles des polymeres semi-conducteurs. J Polym Sci Part C Polym Symp 22:1187–1195
12. Berets DJ, Smith DS (1968) Electrical properties of linear polyacetylene. Trans Faraday Soc 64:823–828
13. Morrissey S (2002) Obituaries. Chem Eng News 80(20):56
14. Anon (2002) Donald Joseph Berets. Harvard Magazine, May–June
15. Berets DJ (1949) Studies on the detonation of explosive gas mixtures. Ph.D. Dissertation, Harvard University
16. Anon (2010) Dorian Sevcik Smith Obituary. Wilmington Star-News, December 15
17. Smith DS (1958) Observations on the rare-earths: chemical and electrochemical studies in non-aqueous solvents. Ph.D. Dissertation. University of Illinois
18. Illinois State Athletics Percy Hall of Fame (1990) 1950 Football. http://goredbirds.com/hof.a spx?hof=281. Accessed 1 Apr 2018
19. State of Illinois (1954) Proceedings of the Teachers College Board of the State of Illinois, July 1, 1953–June 30, 1954, pp 155–156
20. Shirakawa H, Ito T, Ikeda S (1973) Raman scattering and electronic spectra of poly(acetylene). Polym J 4:460–462
21. Ito T, Shirakawa H, Ikeda S (1974) Simultaneous polymerization and formation of poly-acetylene film on the surface of concentrated soluble Ziegler-type catalyst solution. J Polym Sci Polym Chem Ed 12:11–20
22. Shirakawa H, Louis EJ, MacDiarmid AG, Chiang CK, Heeger AJ (1977) Synthesis of electrically conducting organic polymers: Halogen derivatives of polyacetylene, $(CH)_x$. J Chem Soc Chem Commun 578–580
23. Chiang CK, Fincher CR Jr, Park YW, Heeger AJ, Shirakawa H, Louis BJ, Gau SC, MacDiarmid AG (1977) Electrical conductivity in doped polyacetylene. Phys Rev Lett 39:1098–1101
24. MacDiarmid AG, Mikulski CM, Russo PJ, Saran MS, Garito AF, Heeger AJ (1975) Synthesis and structure of the polymeric metal, $(SN)_x$, and its precursor, S_2N_2. J Chem Soc Chem Commun 476–477
25. Mikulski CM, Russo PJ, Saran MS, MacDiarmid AG, Garito AF, Heeger AJ (1975) Synthesis and structure of metallic polymeric sulfur nitride, $(SN)_x$, and its precursor, disulfur dinitride, S_2N_2. J Am Chem Soc 97:6358–6363
26. Chiang CK, Cohen MJ, Garito AF, Heeger AJ, Mikulski CM, MacDiarmid AG (1976) Electrical conductivity of $(SN)_x$. Solid State Commun 18:1451–1455
27. Chiang CK, Cohen MJ, Peebles DL, Heeger AJ, Akhtar M, Kleppinger J, MacDiarmid AG, Milliken J, Moran MJ (1977) Transport and optical properties of polythiazyl bromides: (SNBr 0.4)x. Solid State Commun 23:607–612
28. MacDiarmid AG (2001) Alan G. MacDiarmid. In: Frängsmyr T (ed) Les Prix Nobel. The Nobel prizes 2000. Nobel Foundation, Stockholm, pp 183–190
29. MacDiarmid AG (2005) Oral history interview by Cyrus Mody at University of Pennsylvania, Philadelphia, Pennsylvania. Oral History Transcript #0325. Chemical Heritage Foundation, Philadelphia
30. Hall N (2003) Twenty-five years of conducting polymers. Chem Commun 1–4
31. Hargittai B, Hargittai I (2005) Alan G. MacDiarmid. In: candid science V: conversations with famous scientists. Imperial College Press, London, pp 401–409

32. MacDiarmid AG (1949) Preparation of mono-halogen substituted compounds of sulphur nitride. Nature 164:1131–1132
33. MacDiarmid AG (2001) "Synthetic metals": a novel role for organic polymers (Nobel lecture). Angew Chem Int Ed 40:2581–2590
34. Halford B (2007) Alan MacDiarmid Dies at 79. Chem Eng News 85(7):16
35. MacDiarmid AG (1953) Isotopic Exchange in complex cyanide—simple cyanide systems. Ph.D. Dissertation. University of Wisconsin-Madison
36. MacDiarmid AG (1955) The chemistry of some new derivatives of the silyl radical. Ph.D. Dissertation. University of Cambridge
37. Heeger AJ (2001) Alan J. Heeger. In: Frängsmyr T (ed) Les Prix Nobel. The Nobel prizes 2000. Nobel Foundation, Stockholm, pp 139–143
38. Heeger AJ (2001) Semiconducting and metallic polymers: the fourth generation of polymeric materials (Nobel lecture). Angew Chem Int Ed 40:2591–2611
39. Heeger AJ (2016) Never lose your nerve! World Scientific Publishing, Singapore, pp 23–26
40. Heeger AJ (2016) Never lose your nerve! World Scientific Publishing, Singapore, pp 34–39
41. Heeger AJ (2016) Never lose your nerve!. World Scientific Publishing, Singapore, p 55
42. Heeger AJ (2016) Never lose your nerve! World Scientific Publishing, Singapore, pp 68–71
43. Heeger AJ (2016) Never lose your nerve! World Scientific Publishing, Singapore, pp 77–91
44. Heeger AJ (1961) Studies on the magnetic properties of canted antiferromagnets. Ph.D. Dissertation. University of California, Berkeley
45. Heeger AJ (2016) Never lose your nerve! World Scientific Publishing, Singapore, pp 181–191
46. Heeger AJ (2016) Never lose your nerve! World Scientific Publishing, Singapore, pp 132–150
47. Walatka VV, Labes MM, Perlstein JH (1973) Polysulfur nitride—a one-dimensional chain with a metallic ground state. Phys Rev Lett 31:1139–1142
48. Hsu C, Labes MM (1974) Electrical conductivity of polysulfur nitride. J Chem Phys 61:4640–4645
49. Labes MM, Love P, Nichols LF (1979) Polysulfur nitride-a metallic, superconducting polymer. Chem Rev 79:1–15
50. Mikulski CM, Russo PJ, Saran MS, MacDiarmid AG, Garito AF, Heeger AJ (1975) Synthesis and structure of metallic polymeric sulfur nitride, $(SN)_x$, and its precursor, disulfur dinitride, S_2N_2. J Am Chem Soc 97:6358–6363
51. Mikulski CM, MacDiarmid AG, Garito AF, Heeger AJ (1976) Stability of polymeric sulfur nitride, $(SN)_x$, to air, oxygen, and water vapor. Inorg Chem 15:2943–2945
52. Cohen MJ, Garito AF, Heeger AJ, MacDiarmid AG, Mikulski CM, Saran MS, Kleppinger J (1976) Solid state polymerization of S_2N_2 to $(SN)_x$. J Am Chem Soc 98:3844–3948
53. Rasmussen SC (2016) On the origin of 'synthetic metals'. Mater Today 19:244–245
54. Rasmussen SC (2016) On the origin of "synthetic metals": Herbert McCoy, Alfred Ubbelohde, and the development of metals from nonmetallic elements. Bull Hist Chem 41:64–73
55. Kaner RB, MacDiarmid AG (1988) Plastics that conduct electricity. Sci Am 258(2):106–111
56. Hargittai I (2011) Drive and curiosity: what fuels the passion for science. Prometheus Books, Amherst, pp 173–190
57. Shirakawa H (2001) Hideki Shirakawa. In: Frängsmyr T (ed) Les Prix Nobel. The Nobel prizes 2000. Nobel Foundation, Stockholm, pp 213–216
58. Shirakawa H (2001) The discovery of polyacetylene film: the dawning of an era of con-ducting polymers (Nobel lecture). Angew Chem Int Ed 40:2574–2580
59. Ito T, Shirakawa H, Ikeda S (1975) Thermal cis-trans isomerization and decomposition of polyacetylene. J Polym Sci Polym Chem Ed 13:1943–1950
60. Shirakawa H (2001) The discovery of polyacetylene film: the dawning of an era of con-ducting polymers. In: Frängsmyr T (ed) Les Prix Nobel. The Nobel prizes 2000. Nobel Foundation, Stockholm, pp 217–226
61. Shirakawa H (2001) Nobel lecture: the discovery of polyacetylene film—the dawning of an era of conducting polymers. Rev Modern Phys 73:713–718
62. Shirakawa H (2002) The discovery of polyacetylene film. The dawning of an era of con-ducting polymers. Synth Met 125:3–10

63. Chiang CK, Gau SC, Fincher CR Jr, Park YW, MacDiarmid AG, Heeger AJ (1978) Polyacetylene, $(CH)_x$: n-type and p-type doping and compensation. App Phys Lett 33:18–20
64. Chiang CK, Park YW, Heeger AJ, Shirakawa H, Louis EJ, MacDiarmid AG (1978) Conducting polymers: Halogen doped polyacetylene. J Chem Phys 69:5098–5104
65. Maricq MM, Waugh JS, MacDiarmid AG, Shirakawa H, Heeger AJ (1978) Carbon-13 nuclear magnetic resonance of cis- and *trans*-polyacetylenes. J Am Chem Soc 100:7729–7730
66. Chiang CK, Druy MA, Gau SC, Heeger AJ, Louis EJ, MacDiarmid AG, Park YW, Shirakawa H (1978) Synthesis of highly conducting films of derivatives of polyacetylene, $(CH)_x$. J Am Chem Soc 100:1013–1015
67. Fincher CR Jr, Peebles DL, Heeger AJ, Druy MA, Matsumura Y, MacDiarmid AG, Shirakawa H, Ikeda S (1978) Anisotropic optical properties of pure and doped polyacetylene. Solid State Commun 27:489–494
68. Park YW, Druy MA, Chiang CK, MacDiarmid AG, Heeger AJ, Shirakawa H, Ikeda S (1979) Anisotropic electrical conductivity of partially oriented polyacetylene. J Polym Sci Polym Lett Ed 17:195–201
69. Duke CB, Paton A, Salaneck WR, Thomas HR, Plummer EW, Heeger AJ, MacDiarmid AG (1978) Electronic structure of polyenes and polyacetylene. Chem Phys Lett 59:146–150
70. Goldberg IB, Crowe HR, Newman PR, Heeger AJ, MacDiarmid AG (1979) Electron spin resonance of polyacetylene and AsF_5-doped polyacetylene. J Chem Phys 70:1132–1136
71. Nigrey PJ, MacDiarmid AG, Heeger AJ (1979) Electrochemistry of polyacetylene, $(CH)_x$: electrochemical doping of $(CH)_x$ films to the metallic state. J Chem Soc Chem Commun 594–595
72. Chiang CK, Heeger AJ, Macdiarmid AG (1979) Synthesis, structure, and electrical properties of doped polyacetylene. Ber Bunsen-Ges 78983:407–417
73. Salaneck WR, Thomas HR, Duke CB, Paton A, Plummer EW, Heeger AJ, MacDiarmid AG (1979) J Chem Phys 71:2044–2050
74. Wnek GE, Chien JCW, Karasz FE, Druy MA, Park YW, MacDiarmid AG, Heeger AJ (1979) Variable-density conducting polymers: conductivity and thermopower studies of a new form of polyacetylene: $(CH)_x$. J Polym Sci Polym Lett Ed 17:779–786
75. Weinberger BR, Kaufer J, Heeger AJ, Pron A, MacDiarmid AG (1979) Magnetic susceptibility of doped polyacetylene. Phys Rev B 20:223–230
76. Fincher CE Jr, Ozaki M, Tanaka M, Peebles D, Lauchlan L, Heeger AJ (1979) Electronic structure of polyacetylene: optical and infrared studies of undoped semiconducting $(CH)_x$ and heavily doped metallic $(CH)_x$. Phys Rev B 20:1589–1601
77. Karasz FE, Chien JCW, Galkiewicz R, Wnek GE, Heeger AJ, MacDiarmid AG (1979) Nascent morphology of polyacetylene. Nature 282:286–288
78. Theophilou N, Aznar R, Munardi A, Sledz J, Schue F, Naarnann H (1986) E.S.R. Study of the $Ti(OBu)_4$ catalyst mixture in silicone oil with regard to the synthesis of homogeneous and highly conducting $(CH)_x$. Synth Metals 16:337–342
79. Naarmann H (1990) The development of electrically conducting polymers. Adv Mater 2:345–348
80. Naarmann H (1959) On the identification of constituents of the poison of some bird spiders. Ph.D. Dissertation. Julius-Maximilians-Universität Würzburg
81. Munardi A, Aznar R, Theophilou N, Sledz J, Schue F, Naarnann H (1987) Morphology of polyacetylene produced in the presence of the soluble catalyst $Ti(OnBu)_4$-*n*-BuLi. Eur Polym J 23:11–14
82. Munardi A, Theophilou N, Aznar R, Sledz J, Schue F, Naarnann H (1987) Polymerization of acetylene with $Ti(OC4H9)4$/butyllithium as catalyst system and silicone oil as reaction medium. Makromol Chem 188:395–399
83. Theophilou N, Aznar R, Munardi A, Sledz J, Schue F, Naarnann H (1987) Polymerization of acetylene with new catalytic systems and optimization of the properties of the polymers. J Macromol Sci Chem A24:797–812
84. Naarmann H, Theophilou N (1987) New process for the production of metal-like, stable polyacetylene. Synth Metals 22:1–8

85. Basescu N, Liu ZX, Moses D, Heeger AJ, Naarmann H, Theophilou N (1987) High electrical conductivity in doped polyacetylene. Nature 327:403–405
86. Schimmel T, Rieß W, Gmeiner J, Denninger G, Schwoerer M, Naarmann H, Theophilou N (1988) DC-Conductivity on a new type of highly conducting polyacetylene, N-(CH)$_x$. Solid State Commun 65:1311–1315
87. Naarmann H, Theophilou N (1989) Influences of the catalyst system on the morphology, structure and conductivity of a new type of polyacetylene. Makromol Chem Macromol Symp 24:115–128
88. Schimmel T, Denninger G, Riess W, Voit J, Schwoerer M, Schoepe W, Naarmann H (1989) High-σ polyacetylene: DC conductivity between 14 mK and 300 K. Synth Metals 28:D11–D18
89. Winter H, Sachs G, Dormann E, Cosmo R, Naarmann H (1990) Magnetic properties of spin-labelled polyacetylene. Synth Metals 36:353–365
90. Schimmel Th, Schwoerer M, Naarmann H (1990) Mechanisms limiting the D.C. conductivity of high-conductivity polyacetylene. Synth Metals 37:1–6
91. Schimmel T, Glaser M, Schwoerer M, Naarmann H (1991) Conductivity barriers and transmission electron microscopy on highly conducting polyacetylene. Synth Metels 41–43:19–25

Chapter 7
2000 Nobel Prize in Chemistry

The awarding of the Nobel Prize in Chemistry to Hideki Shirakawa, Alan MacDiarmid and Alan Heeger in 2000 [1] elevated the awareness of polyacetylene among the general science community and cemented the place of this polymeric material in the history of chemistry. Although acetylene polymers will always be most closely associated with the 2000 Nobel Prize in Chemistry, the previous chapters bring to light that it was only the latest award in a series of Nobel Prizes connected to polymers of acetylene. Paul Sabatier was awarded the 1912 Nobel Prize in Chemistry[1] *"for his method of hydrogenating organic compounds in the presence of finely disintegrated metals"* [2], work that led directly to his production of cuprene via the polymerization of acetylene over copper powder in 1899.[2] This was then followed by the awarding of the 1938 Nobel Prize in Chemistry to Richard Kuhn *"for his work on carotenoids and vitamins"* [3], which included his pioneering work on polyenes as models of these systems. Of course, these same polyenes also became models for conjugated polyacetylene.[3] Lastly, Giulio Natta shared the Nobel Prize in Chemistry with Karl Ziegler in 1963 *"for their discoveries in the field of the chemistry and technology of high polymers"* [4]. Among these stated discoveries of Natta was, of course, the very first successful production of polyacetylene in 1955.[4] Still, while these previous awards are at least partially connected to acetylene polymers, the 2000 Nobel Prize is the only one directly given for work on a polymer of acetylene and thus this final chapter will conclude with a discussion of the awarding of this prize.

---

[1] Shared with Victor Grignard "for the discovery of the so-called Grignard reagent, which in recent years has greatly advanced the progress of organic chemistry" [2].

[2] See Chap. 3.

[3] See Chap. 4.

[4] See Chap. 5.

© The Author(s) 2018
S. C. Rasmussen, *Acetylene and Its Polymers*, SpringerBriefs in Molecular Science,
https://doi.org/10.1007/978-3-319-95489-9_7

7.1 Details of the 2000 Nobel Prize in Chemistry

On October 10, 2000, The Royal Swedish Academy of Sciences announced via a press conference and associated press release [5] that the Nobel Prize in Chemistry for 2000 was to be awarded jointly to Alan J. Heeger, Alan G. MacDiarmid, and Hideki Shirakawa *"for the discovery and development of conductive polymers."* The prize included a monetary amount of 9 million Swedish Krona,[5] which would also be shared equally among the three awardees. In the accompanying *Advanced Information* released by the Academy, it is stated that [6]:

> The choice is motivated by the important scientific position that the field has achieved and the consequences in terms of practical applications and of interdisciplinary development between chemistry and physics.

The Swedish Nobel Committee for Chemistry in 2000 was made up of nine professors from seven different Swedish universities (Table 7.1), although all were from only four cities in Sweden [7]. Bengt Nordén (b. 1945) of Chalmers University of Technology served as Chairman of the Committee, with Astrid Graslund (b. 1945) of Stockholm University serving as the Secretary of the Committee [6, 7]. The fact that over half the committee were from fields with strong ties to physics (i.e. physical chemistry, theoretical chemistry, biophysics) most likely played a significant role in the stated motivation concerning the interdisciplinary development between chemistry and physics and the awarding of the prize for a topic shared equally between chemistry and physics.

The awardees were notified by the President of the Swedish Academy of Sciences via phone on October 10, 2000, shortly before the official announcement was made at 3:15 pm in Stockholm during a press conference. According to Heeger, he received his call at 5:45 am in Santa Barbara, 30 min before the Stockholm press conference [8]. The formal awarding of the Nobel Prize medals and diplomas occurred during ceremonies at the Stockholm Concert Hall in Sweden on December 10, 2000 [9]. As part of the ceremonies, Bengt Nordén gave an introductory speech on the 2000 Nobel Prize in Chemistry [10], after which the Nobel laureates received their prize from Carl XVI Gustaf, the King of Sweden [8]. Following the ceremony itself was the Nobel Banquet held on the lower floor of the City Hall, during which all three Nobel laureates gave a brief speech.

Prior to the award ceremonies, the three laureates also gave Nobel Lectures on December 8, 2000 at Stockholm University [11–13]. Alan Heeger's lecture was entitled "Semiconducting and Metallic Polymers: The Fourth Generation of Polymeric Materials" [11], with Alan MacDiarmid's entitled" "Synthetic Metals": A Novel Role for Organic Polymers" [12] and Hediki Shirakawa's entitled "The Discovery of Polyacetylene Film: The Dawning of an Era of Conducting Polymers" [13]. Various versions of these lectures were then reprinted in the journals *Angewandte Chemie International Edition* [14–16], *Synthetic Metals* [17–19], and *Reviews of Modern Physics* [20].

---

[5]In 2000, this was equal to ca. $982,300 US.

Table 7.1 Members of the 2000 Swedish nobel committee for chemistry [7]

Name	Field	University	City
Bengt Nordén[a]	Physical chemistry	Chalmers University	Gothenburg
Björn Roos	Theoretical chemistry	Lund University	Lund
Carl-Ivar Brändén	Molecular biology	Karolinska Institute	Stockholm
Ingmar Grenthe	Inorganic chemistry	Royal Institute of Technology	Stockholm
Per Ahlberg	Organic chemistry	University of Gothenburg	Gothenburg
Astrid Graslund[b]	Biophysics	Stockholm University	Stockholm
Torvard Laurent[c]	Medical and physiological chemistry	Uppsala University	Uppsala
Hakan Wennerström[c]	Theoretical physical chemistry	Lund University	Lund
Gunnar von Heijne[c]	Theoretical chemistry	Stockholm University	Stockholm

[a]Chairman of the Committee, [b]Secretary of the Committee, [c]Adjoint member

7.2   What Is Discovery?

One point of debate concerning the 2000 Nobel Prize in Chemistry has been the official wording that specifies Heeger, MacDiarmid, and Shirakawa as the discoverers of conductive polymers. In his Nobel Lecture, MacDiarmid tried to clarify that the class of polymers recognized by the award are different then those described by the more general description of "conducting polymers" [12]:

> This class of polymer is completely different from "conducting polymers" which are merely a physical mixture of a nonconductive polymer with a conducting material such as a metal or carbon powder distributed throughout the material.

In contrast, the "intrinsically conducting polymers" described by MacDiarmid are conjugated organic polymers whose conductivity are enhanced through the process of doping, such as that demonstrated for polyacetylene in the work of the three Nobel laureates beginning in 1977 [21]. However, in documenting the history of these materials [22–28], it has been highlighted that published studies on such conjugated polymers date back to the 1834 work of F. Ferdinand Runge (1794–1867) [29], with reports of the conductive nature of doped polymers beginning in 1963 with the work of Donald Weiss (1924–2008) in Australia [30–32]. Of course, one could argue that the conductivities reported by Wiess were still quite low (ca. 0.1 S cm^{-1}) [32] and Wiess himself states that he did not fully understand the role of doping on the polymer conductivity at the time [22, 24]. The same could not be said, however, of the work of Rene Buvet (1930–1992) and Marcel Jozefowicz (b. 1934) on polyaniline beginning in 1966 [33–36], who had achieved conductivities as high as 100 S cm^{-1} by 1969 [36].

Even looking to the secondary literature of the 1960s, several reviews of conductive polymers were published prior to the 1977 polyacetylene work, with the first such review published in 1965 [37–42]. To add to these works, three books entitled *Organic Semiconductors* were published during the 1960s, all of which included conductive polymers [43–45]. Finally, a research report entitled *Organic Semiconductors—Their Technological Promise* was also published in 1962 by the United States Department of Commerce [46]. While much of this secondary literature focused more on examples of what MacDiarmid defined as conducting polymers, conjugated polymers were also included as well.

Looking to reviews on conductive polymers published in the initial years following the polyacetylene work of Heeger, MacDiarmid, and Shirakawa, it is quite clear that doped polyacetylenes were now included, but these reviews do not portray the doped polymers as unique and are given as just one of many examples of conductive materials [47, 48]. As such, the polyacetylene results are not presented as particularly revolutionary or the clear start of a new field. In fact, a 1982 review by Karlheinz Seeger (1927–2008) of the University of Vienna states [48]:

> All polymers to be discussed below are in fact well known for many years or even decades. Also the technique of increasing the conductivity by doping with halogens or sulfur is not new. It was not known, however, that a metallic conductivity would be obtained at high doping levels.

When viewed in relation to this wealth of previous work, it is hard to understand the committee's view to designate the Nobel laureates as the discoverers of these materials. One must question if they were unaware of this previous work or if there was some specific achievement in the work of Heeger, MacDiarmid, and Shirakawa that they felt justified the stated designation. In reviewing the *Advanced Information* released by the Academy [6], it is stated *"However, polyacetylene was the conductive polymer that actually launched this new field of research,"* followed by reference to two reviews on the field [49, 50]. Although the review by Kanatzidis [49] does provide a narrative that the field began with polyacetylene, the much more comprehensive and historically accurate review by Feast and Meijer [50] does not and only states that the work of Heeger, MacDiarmid, and Shirakawa marked a new phase in the study of polyacetylene. More so, this more comprehensive review does accurately cover much of the early history of the field as outlined above. As such, the committee must have been aware of this previous work.

In light of this discussion, one then must ask what is actually meant by "discovery". Of course, the complex nature of scientific discovery has been a recurring subject in the history and philosophy of science [51–54]. For many, the act of discovery is what is often described as the "eureka moment" [51, 52]. As highlighted by the *Stanford Encyclopedia of Philosophy* [51], the act of having an insight, the "eureka moment," was distinguished from the later processes of articulating, developing, and testing that insight during the course of the 19th century. An important contribution to debates about scientific discovery was the work of William Whewell (1794–1866), as he clearly separated the eureka moment, or "happy thought", from other elements of scientific inquiry. For him, the complete process of discovery consisted of three elements: the happy thought, the articulation of that thought, and its testing or verification [51]. The common view of discovery, however, is often limited to either the eureka moment alone, or to just the eureka moment and its articulation. Still, for others such as the physicist and philosopher of science Thomas Kuhn (1922–1996), discovery is not considered a simple act, but an extended, complex process, which ultimately results in a paradigm shift [51]. In this light, a discovery "normally entails the abandonment, or at least the qualification, of earlier theories or observations" [53]. Of course, this can complicate the narrative of popular interpretations of scientific history, which are often biased towards a single great discovery by select figures of great stature, particularly events easily commemorated in an anniversary. Most discoveries, however, are much more nuanced, complicated, and communal [54].

When asked during an interview about the nature of the contributions of Pyun and Shirakawa on the discovery of polyacetylene films, MacDiarmid stated [55]:

> Of course, your remark touches an interesting aspect of this discovery and, generally speaking, of scientific discoveries. Who is the discoverer? Is it the person who does something mechanically or is it the person who realizes its significance? Quite often the person who does the mechanical operation in the lab is also the person that realizes its significance. Sometimes it is not the same person.

Still, this then leaves the lingering question concerning what aspect of the polyacetylene work of the 1970s that the Nobel committee felt went far beyond or overturned earlier theories or observations. The highlighted polyacetylene work definitely achieved much higher conductivities than previous efforts, and ultimately resulted in a greater understanding of the doping process, yet it is difficult to pinpoint a resulting paradigm shift in the development of conductive polymers via the doping of conjugated materials. In 2005, Heeger contributed his opinion on the topic with the brief comment [56]:

> We were the pioneers. This is not to say that many other people didn't make important contributions.

More recently, however, he expanded on this issue of discovery in his 2016 autobiography, stating [57]:

> Had we really created this new field? Were we really first? The answer to such questions was clearly and eloquently first stated by Sir Isaac Newton: "If I have seen further, it is by standing on the shoulders of giants."[6] Many chemist and physicists had touched upon the subject of conjugated polymers. There were reports in the literature of modest electrical conductivities. We were aware of some of these and ignorant of others. I will not attempt to list those earlier "giants" because I would surely leave out important names that should have been included. Our interdisciplinary approach, breath and solidity of our results, the new theoretical concepts that we introduced, and the firm scientific foundation that we build all contributed to what turned out to be the creation of the field of semiconducting and metallic polymers.

---

[6]Although the most familiar expression of this statement in English is by Isaac Newton in 1675, the origin of the statement has been traced to the 12th century and attributed to Bernard of Chartres [58].

As can be seen, it is not possible to come to a simple conclusion regarding either the nature of discovery or attribution of discovery in relation to conducting polymers. In the end, the overall complicated debate on scientific discovery is perhaps best summed up by the following editorial statement in *Nature* [54]:

> Such is the nature of discovery — incremental at times, fast-paced at others, occasionally derailing into pettiness.

7.3   Legacy

No matter one's views on the debate of Heeger, MacDiarmid, and Shirakawa's "discovery" of conductive polymers, there is no doubt that their work was a pivotal point in the history of conjugated polymers. Not only were they the first to demonstrate conductive polymers with truly metallic conductivities, but the visual appeal of an organic plastic that both looked and acted metallic helped attract a greater number of intrigued scientists to study these conjugated materials. As a result, this helped to build a critical mass to drive innovation forward. Perhaps of greatest importance, their work marked the point in the evolution of conductive polymers in which separate lines of investigation in both chemistry (conjugated polymers) and physics (conductive organic solids) really came together for the first time, thus resulting in the current interdisciplinary nature of the field. In the end, the field of conjugated materials owes much to Heeger, MacDiarmid, and Shirakawa, whose work initiated a new phase in the study of these materials and sparked the rapid growth of a niche area of scientific interest into the wide community of modern organic electronics. In addition, their contributions did not end with their initial work on polyacetylene, with MacDiarmid and Heeger remaining central figures in the further growth of the field for decades afterwards. As such, it was with great joy that the field received the news in the fall of 2000, that Heeger, MacDiarmid, and Shirakawa were being justly recognized by the Nobel committee for their contributions to the field of conjugated and conducting polymers. Still, it is important for all of us to understand that this field was not suddenly born in the 1970s and to recognize its very deep history that spans centuries. Not only does this properly credit all of the important early contributions that helped shape the origins of the field, but also reveals that these pursuits make up a more significant part of the global histories of chemistry and physics than is often recognized.

References

1. Nobelprize.org. Nobel Media AB (2014) The nobel prize in chemistry 2000. http://www.nobe lprize.org/nobel_prizes/chemistry/laureates/2000/. Accessed 10 May 2018
2. Nobelprize.org. Nobel Media AB (2014) The nobel prize in chemistry 1912. http://www.nobe lprize.org/nobel_prizes/chemistry/laureates/1912/. Accessed 10 May 2018

3. Nobelprize.org. Nobel Media AB (2014) The nobel prize in chemistry 1938. http://www.nobe lprize.org/nobel_prizes/chemistry/laureates/1938/. Accessed 10 May 2018
4. Nobelprize.org. Nobel Media AB (2014) The nobel prize in chemistry 1963. http://www.nobe lprize.org/nobel_prizes/chemistry/laureates/1963/. Accessed 10 May 2018
5. Nobelprize.org. Nobel Media AB (2014) Press release: the 2000 nobel prize in chemistry. http://www.nobelprize.org/nobel_prizes/chemistry/laureates/2000/press.html. Accessed 10 May 2018
6. Nobelprize.org. Nobel Media AB (2014) The nobel prize in chemistry 2000—advanced information. http://www.nobelprize.org/nobel_prizes/chemistry/laureates/2000/advan ced.html. Accessed 10 May 2018
7. Frängsmyr T (ed) (2001) Les Prix Nobel. The nobel prizes 2000, Nobel Foundation, Stockholm, p 8
8. Heeger AJ (2016) Never lose your nerve! World Scientific Publishing, Singapore, pp 151–163
9. Nobelprize.org. Nobel Media AB (2014) The nobel prize award ceremony 2000. http://www.no belprize.org/nobel_prizes/chemistry/laureates/2000/award-video.html. Accessed 10 May 2018
10. Nordén B (2001) The nobel prize in chemistry. In: Frängsmyr T (ed) Les Prix Nobel. The nobel prizes 2000, Nobel Foundation, Stockholm, pp 21–23
11. Heeger AJ (2001) Semiconducting and metallic polymers: the fourth generation of polymeric materials. In: Frängsmyr T (ed) Les Prix Nobel. The nobel prizes 2000, Nobel Foundation, Stockholm, pp 144–181
12. MacDiarmid AG (2001) "Synthetic metals": a novel role for organic polymers. In: Frängsmyr T (ed) Les Prix Nobel. The nobel prizes 2000, Nobel Foundation, Stockholm, pp 191–211
13. Shirakawa H (2001) The discovery of polyacetylene film: the dawning of an era of conducting polymers. In: Frängsmyr T (ed) Les Prix Nobel. The nobel prizes 2000, Nobel Foundation, Stockholm, pp 217–266
14. Shirakawa H (2001) The discovery of polyacetylene film: the dawning of an era of conducting polymers (Nobel Lecture). Angew Chem Int Ed 40:2574–2580
15. MacDiarmid AG (2001) "Synthetic metals": a novel role for organic polymers (Nobel Lecture). Angew Chem Int Ed 40:2581–2590
16. Heeger AJ (2001) Semiconducting and metallic polymers: the fourth generation of polymeric materials (Nobel Lecture). Angew Chem Int Ed 40:2591–2611
17. Shirakawa H (2002) The discovery of polyacetylene film. The dawning of an era of con-ducting polymers. Synth Met 125:3–10
18. MacDiarmid AG (2002) Synthetic metals: a novel role for organic polymers. Synth Met 125:11–22
19. Heeger AJ (2002) Semiconducting and metallic polymers: the fourth generation of polymeric materials. Synth Met 125:23–42
20. Shirakawa H (2001) Nobel lecture: The discovery of polyacetylene film—the dawning of an era of conducting polymers. Rev Modern Phys 73:713–718
21. Shirakawa H, Louis EJ, MacDiarmid AG, Chiang CK, Heeger AJ (1977) Synthesis of electrically conducting organic polymers: halogen derivatives of polyacetylene, $(CH)_x$. J Chem Soc Chem Commun 578–580
22. Rasmussen SC (2011) Electrically conducting plastics: revising the history of conjugated organic polymers. In: Strom ET, Rasmussen SC (eds) 100+years of plastics: Leo Baekeland and beyond, acs symposium series 1080. American Chemical Society, Washington, DC, pp 147–163
23. Rasmussen SC (2014) The path to conductive polyacetylene. Bull Hist Chem 39:64–72
24. Rasmussen SC (2015) Early history of polypyrrole: the first conducting organic polymer. Bull Hist Chem 40:45–55
25. Rasmussen SC (2016) On the origin of 'synthetic metals'. Mater Today 19:244–245
26. Rasmussen SC (2016) On the origin of "synthetic metals": Herbert McCoy, Alfred Ubbelohde, and the development of metals from nonmetallic elements. Bull Hist Chem 41:64–73
27. Rasmussen SC (2017) Early history of conductive organic polymers. In: Zhang Z, Rouabhia M, Moulton SE (eds) Conductive polymers: electrical interactions in cell biology and medicine. CRC Press, Boca Raton, FL, 2017; Chapter 1

28. Rasmussen SC (2017) The early history of polyaniline: discovery and origins. Substantia 1(2):99–109
29. Runge FF (1834) Ueber einige Producte der Steinkohlen-destillation. Ann Phys Chem 31:513–524
30. McNeill R, Siudak R, Wardlaw JH, Weiss DE (1963) Electronic conduction in polymers. Aust J Chem 16:1056–1075
31. Bolto BA, Weiss DE (1963) Electronic conduction in polymers. II. The electrochemical reduction of polypyrrole at controlled potential. Aust J Chem 16:1076–1089
32. Bolto BA, McNeill R, Weiss DE (1963) Electronic conduction in polymers. III. Electronic properties of polypyrrole. Aust J Chem 16:1090–1103
33. Jozefowicz M, Yu LT (1966) Relations entre propriétés chimiques et électrochimiques de semi-conducteurs macromoléculaires. Rev Gen Electr 75:1008–1013
34. Yu LT, Jozefowicz M (1966) Conductivité et constitution chimique pe semi-conducteurs macro-moléculaires. Rev Gen Electr 75:1014–1018
35. De Surville R, Jozefowicz M, Yu LT, Perichon J, Buvet R (1968) Electrochemical chains using protolytic organic semiconductors. Electrochim Acta 13:1451–1458
36. Jozefowicz M, Yu LT, Perichon J, Buvet R (1969) Proprietes Nouvelles des Polymeres Semi-conducteurs. J Polym Sci Part C Polym Symp 22:1187–1195
37. Weiss DE, Bolto BA (1965) Organic polymers that conduct electricity. In: Physics and chemistry of the organic solid state. Interscience Publishers, New York, vol II, Chapter 2
38. Labes MM (1966) Conductivity in polymeric solids. Pure Appl Chem 21:275–285
39. Trivedi PD (1968) Electrically conductive polymers. Pop Plast 13(9):25–9; 13(10): 30–5
40. Rembaum A (1969) Conductive polymers. Encycl Polym. Sci Technol 11:318–337
41. Lupinski JH (1969) Conductive polymers. Ann N Y Acad Sci 155(2):561–565
42. Goodings EP (1970) Conductive polymers. Rep Prog Appl Chem 55:53–65
43. Brophy JJ, Buttrey JW (eds) (1962) Organic semiconductors. Proceedings of an interindustry conference. The Macmillan Company, New York
44. Okamoto Y, Brenner W (1964) Organic semiconductors. Rheinhold, New York
45. Gutmann F, Lyons LE (1967) Organic semiconductors. Wiley, New York
46. Office of Technical Services, U.S. Department of Commerce (1962) Organic semiconductors—their technological promise. U.S. Government Research Report, PB 181037
47. Mort J (1980) Conductive polymers. Science 208:819–825
48. Seeger K (1982) The morphology and structure of highly conducting polymers. Angew Makromol Chem 109(110):227–251
49. Kanatzidis MG (1990) Conductive polymers. Chem Eng News 68(49):36–54
50. Feast WJ, Tsibouklis J, Pouwer KL, Groenendaal L, Meijer EW (1996) Synthesis, processing and material properties of conjugated polymers. Polymer 37:5017–5047
51. Schickore J (2014) Scientific discovery. In: Zalta EN (ed) The Stanford encyclopedia of philosophy. https://plato.stanford.edu/archives/spr2014/entries/scientific-discovery/. Accessed 10 May 2018
52. Hargittai I (2011) Risking reputation: conducting polymers. Drive and curiosity: what fuels the passion for science. Prometheus Books, Amherst, NY, pp 173–190
53. Fox R (2014) The nature of discovery. Notes Rec 68:319–321
54. Anon (2017) Awkward first dates. Nature 550:7
55. Hargittai B, Hargittai I (2005) Alan G. MacDiarmid. In: Candid science V: conversations with famous scientists. Imperial College Press, London, pp 401–409
56. Hargittai B, Hargittai I (2005) Alan J. Heeger. In: Candid science V: conversations with famous scientists. Imperial College Press, London, pp 411–427
57. Heeger AJ (2016) Never lose your nerve! World Scientific Publishing, Singapore, pp 143–144
58. MacGarry DD (ed) (1955) The metalogicon of John Salisbury: A twelfth-century defense of the verbal and logical arts of the Trivium. University of California Press, Berkeley, p 167

Index

© The Author(s) 2018
S. C. Rasmussen, *Acetylene and Its Polymers*, SpringerBriefs in Molecular Science,
https://doi.org/10.1007/978-3-319-95489-9

Printed in the United States
By Bookmasters